IT'S ABOUT TIME

IT'S ABOUT TIME

UNDERSTANDING EINSTEIN'S RELATIVITY

N. David Mermin

PRINCETON UNIVERSITY PRESS PRINCETON AND OXFORD

Published by Princeton University Press, 41 William Street, Princeton, New Jersey 08540
In the United Kingdom: Princeton University Press, 6 Oxford Street, Woodstock,
Oxfordshire OX20 1TW

Sixth printing, and first paperback printing, 2009
Paperback ISBN: 978-0-691-14127-5

The Library of Congress has cataloged the cloth edition of this book as follows

Mermin, N. David.
  It's about time : understanding Einstein's relativity / N. David Mermin.
    p. cm.
  Includes index.
  ISBN-13: 978-0-691-12201-4 (cloth : alk. paper)
  ISBN-10: 0-691-12201-6 (cloth : alk. paper)
  1. Special relativity (Physics) I. Title.

QC173.65.M47 2005
530.11–dc22
                2004042068

British Library Cataloging-in-Publication Data is available

This book has been composed in Sabon

Printed on acid-free paper. ∞

press.princeton.edu

Printed in the United States of America

10   9

for Hannah and Sam

At last it came to me that time was suspect!
—*Albert Einstein*

Contents

Preface: Why Another Relativity Book _____

Absolute, true, and mathematical time, of
itself and from its own nature, flows
equably without relation to anything
external.
    —*Isaac Newton*

My time is your time.
    —*Rudy Vallee*

It came to me that time was suspect!
    —*Albert Einstein*

The year 2005 is the centenary of the publication of Einstein's special
theory of relativity. Forty years earlier, in the 60th anniversary year, as
a new young assistant professor of physics at Cornell, I decided that it
was about time to make relativity a standard part of the high school curriculum. This can be done by incorporating it into courses in elementary
algebra or plane geometry, which are, surprisingly, the only technical tools
required for a full understanding of the subject. So I taught a course on
special relativity to a group of high school teachers, and they seemed to
enjoy it.

Relativity is perfect for the high school curriculum not only because it
offers an astonishing application of elementary high school mathematics,
but also because everybody is intimately acquainted with the subject of
relativity. Relativity is about time. What could be more familiar? What
makes the subject so fascinating is that relativity reveals the nature of
time to be shockingly different from what had been taken completely
for granted, up until 1905. We now know, for example, that the first
two statements that open this preface are incorrect. Understanding why
Newton and Vallee were both wrong about time ought to be an important
part of anybody's education. The wonderfully succinct characterization,
in the third statement, of the key to a mystery facing physicists at the
start of the 20th century is from a report of a private conversation with
Einstein, toward the end of his life.[1]

_____

[1] R. S. Shankland, "Conversations with Albert Einstein," *American Journal of Physics*
31 (1963): 47–57.

In spite of these good reasons, during the past forty years special rela-
tivity has not been incorporated into the standard high school curriculum.
The only thing to emerge from my quixotic effort was a book in 1968,
Space and Time in Special Relativity, which remains in print to this day.
Although I thought I was writing it for high school students, during the
three and a half decades since its appearance, it has rarely showed up in
high schools. Instead I have used it in courses on special relativity that
I offered intermittently to nonscientists at Cornell over the past three
decades.

In the course of explaining the subject to students over many years,
I found I was becoming increasingly dissatisfied with my own book.
Although I preferred it to any of the other treatments available at its elementary mathematical level, I came more and more to view it as the best
of an unsatisfactory bunch. By the 1990s I had ceased to recommend it as
a text, relying instead on a set of lecture notes I was writing specifically for
my Cornell nonscientists. Throughout that decade these notes remained
a work in progress, being continually reorganized and revised in response
to newly discovered difficulties and misunderstandings that emerged in
the course of innumerable conversations with smart, puzzled students.

The visionary assistant professor of forty years ago is now about to
retire, and this new book on special relativity grew out of the current
form of my lecture notes. I have no doubt that the notes would continue
to improve were I to continue to have the stimulation of the wonderfully
bright, outspoken, skeptical Cornell students who helped them evolve into
their present form. But without guidance from that source of inspiration
and surprises, further tinkering is as likely to make my notes worse as
better. It's time to transform them into a new book.

Between 1968 and 2005 I've learned a lot about explaining special relativity. One pedagogical discovery has been especially valuable. Anybody
wishing to understand the subject must be able to visualize how certain
events taking place, say, in a railroad station, are described from the point
of view of a passenger passing through that station on a uniformly moving
train and, conversely, how events taking place on such a train appear to a
person standing in the station. Without the ability to translate from one
such description to another, one cannot begin to understand relativity.
But all introductions to relativity that I know of, including my own 1968
book, take the ability to do this for granted. They immediately require
the reader to apply this unused, undeveloped, often nonexistent skill to
some highly counterintuitive phenomena.

In explaining relativity this process often leads to descriptions from two
different perspectives, which appear, at first glance, to contradict each
other. Faced with an apparent paradox, people who have never before
thought about transforming station descriptions to train descriptions and

vice versa quite reasonably assume that they must have done something wrong in the transcription. Rather than seeking an understanding of why the contradiction is only apparent, they lose confidence in the analytical technique that gave rise to it.

In this respect the pedagogy of the standard approach to relativity is terrible. One introduces a crucial and unfamiliar conceptual technique—changing descriptions from one "frame of reference" to another—by immediately applying it to some unusual and highly counterintuitive cases. The most important thing I learned in teaching relativity to many generations of Cornell undergraduates, none of them science majors, is that one must begin teaching them the technique of changing frames of reference by applying that technique to some entirely commonplace, highly intuitive examples. There are many such ways to develop these skills, and they enable one to learn much that is not at all obvious, though never paradoxical. This is the subject of chapter 1 of the present book, where we examine some simple questions about colliding objects. Although one rarely if ever talks about such "nonrelativistic" phenomena as a preliminary to conventional expositions of relativity, I am now convinced that it is essential to do so in explaining the subject to people with no formal scientific background, if they are to acquire any under-standing of what follows. Beginning an introduction to relativity with an examination of simple collisions also has a secondary benefit, because they are then available to play quite a different role, in the explanation of $E = mc^2$.

Another thing I have learned since 1968 is that one should emphasize as early as possible that although objects moving at the speed of light famously behave in some very strange ways, the behavior of objects mov-ing at speeds comparable to the speed of light can be just as peculiar. The peculiarity of motion at the speed of light is just a special case of a more general peculiarity of all motion, which becomes prominent only at extremely high speeds. That more general peculiarity can be expressed by an elementary but precise rule that it is possible and useful to formulate at a very early stage of the subject. I do this in chapter 4, using a surprisingly simple thought experiment, which appeared as a homework problem in my 1968 book. When I realized that nobody seemed ever to have noticed this argument, I published my homework problem (and its solution) in the *American Journal of Physics* (1983). Subsequently, it became clear to me that the subject of the homework problem has an important ped-agogical role to play in the development of relativity. It enables one to see that many tricks with light that seem to give it a fundamental role in establishing the nature of time can, in fact, be done just as well using any other uniformly moving things to signal from one place to another.

A less unorthodox feature of my current presentation of relativity is one whose importance Einstein understood from the beginning, but which some later treatments of the subject, including my own 1968 book, have tended to underemphasize. This is the entirely conventional nature of the claim that two events that happen in different places are simultaneous. That no inherent meaning can be assigned to the simultaneity of distant events is the single most important lesson to be learned from relativity, and it is an unavoidable component of any introduction to the subject. But in 1968 I introduced it as a secondary consequence of some other peculiarities, rather than insisting on it as the crucial key to making sense of just about everything else. In the present volume the conventional nature of simultaneity is introduced at a very early stage, and its quantitative content is given a simple, concise, easily remembered formulation that plays a central role in clarifying everything that follows.

Another innovation in the present book is my treatment of the space-time diagrams invented by Minkowski in the early days of the subject. These play an important role in tying everything together in an intuitive set of pictures, uncomplicated by equations. In my 1968 book these diagrams appear in a fairly conventional setting, in which space-time coordinate axes play a fundamental role in the description of events, and (less conventionally) trigonometric identities are used to extract important information. Some twenty-five years after that, I realized that the axes are an unnecessary and potentially confusing distraction, and that all of the sometimes cumbersome trigonometry of my earlier exposition can be replaced by some very simple plane geometry, generally involving little more than the identification of various sets of similar triangles. As far as I know, this approach to space-time diagrams, which extracts them with a minimum of analysis directly from Einstein's two principles, has never before appeared in a book, or even in the scientific literature until I reported it a few years ago in the *American Journal of Physics* (1997 and 1998). Space-time diagrams, as I present them here, are to conventionally presented space-time diagrams as the plane geometry of Euclid is to the analytic geometry of Descartes. Analytic geometry is the more powerful tool for professional calculations; Euclid's approach is essential for a deeper understanding.

An unusual feature of my 1968 book was an alternative pictorial approach to subject, less powerful, but also less abstract than the space-time diagrams, based on a couple of pictures of two trains in relative motion, showing both trains as they would appear to the passengers on one of them. A decade later it dawned on me that this exposition could be made much simpler and more powerful by showing both trains not from the perspective of either one of them, but from the point of view of a platform along which they moved in opposite directions at the same speed. The exposition of this better approach is the subject of chapter 9.

Finally, in listing the refinements in relativistic pedagogy embodied in the present work, I must mention Alice and Bob. They have been the protagonists of the tales cryptographers like to tell each other for many decades. I made their acquaintance in the 1990s when I became interested in some remarkable new developments in the application of quantum physics to information processing, and I realized that they had an important new role to play in the exposition of special relativity. It's not just that "Alice" and "Bob" are much more pleasant to talk about than "Frame [of reference] A" and "Frame B." The fact that each of them comes equipped with a distinct set of pronouns—at least in European languages—makes some otherwise quite cumbersome narratives entirely colloquial and informal, without sacrificing any precision. They (and occasionally their friends Charles, Carol, Dick, and Eve) play central roles in the present volume, and if this book makes no other contribution to spreading the public understanding of relativity, I hope it will at least pave the way for more appearances of Alice and Bob in the relativistic arena.

Note to Readers

Although the mathematical level of the book is elementary—simple plane geometry and beginning high school algebra—it cannot be read like a novel. I have scrupulously tried to adhere to the rule (widely attributed to Einstein) that the exposition should be as simple as possible, but no simpler than that. It will therefore often be necessary for you to pause and reflect on a line of argument, to examine a figure and contemplate its relation to the accompanying text, and, in general, to participate actively in the process of thinking things through, rather than passively reading along.

While writing the book, I had in mind readers with a range of backgrounds. I have written primarily for people with no training in physics, and none in mathematics beyond very elementary geometry and algebra. But because my approach to the subject contains novel elements and is more down-to-earth and intuitive than the comparatively abstract treatments from which physical scientists usually learn the subject, I would expect that undergraduate physics majors, graduate students, and even professional practitioners of relativity might find a few interesting things here and there in spite of the scrupulously elementary level.

Because of this secondary audience, there are places toward the end of several chapters where I delve a bit further into the subject than would be strictly necessary for my primary readership, though still at the same elementary level. If, as you get deeper into a chapter, you find that the going is getting a little rough, I would urge you to move on to the start of the next chapter. It may well be that the material you skipped over, while of some interest in its own right, plays no role in the development that follows. If this turns out not to be so—if you run across important references back to sections you had bypassed—then, and only then, you might go back and take another run at getting through the relevant material. So while I would not recommend randomly dipping into the book, as one might approach a user's manual, neither is it necessary to read it sequentially from cover to cover.

As examples of such "optional" subjects that might strain the patience of the general reader, I would mention the applications of the relativistic velocity addition law to collisions at the end of chapter 4, the determination of simultaneous events using signals slower than the speed of light at the end of chapter 5, the method of clock synchronization by direct transport of clocks at the end of chapter 6, the discussion at the end of chapter 10 of how the scale factors for different frames are related and the implications of this for the invariant interval, and the various applications of the relativistic conservation laws at the end of chapter 11. I have separated these sections from the rest of their chapters by a paragraph in small print summarizing their contents. While they can be passed over without loss of continuity in the overall argument of the book, I do not encourage you to ignore them, because they contain some very beautiful aspects of relativity, treated in the simplest way I know. But the fact is they can be omitted, without diminishing the understanding of time that the theory reveals.

One

The Principle of Relativity

THE SPECIAL THEORY OF RELATIVITY was set forth by Einstein in his 1905 paper "On the Electrodynamics of Moving Bodies."[1] The term "special relativity" is used to distinguish the theory from Einstein's theory of gravity, known as general relativity, which he completed ten years later. Except for a glimpse into general relativity in chapter 12, we shall be concerned entirely with special relativity, so from now on I will drop the "special," with the understanding that "relativity" always refers to special relativity.

Einstein based the theory of relativity on two postulates. The first is now known as the principle of relativity. We shall take up the second in chapter 3. Einstein put the principle of relativity this way: "In electromagnetism as well as in mechanics, phenomena have no properties corresponding to the concept of absolute rest." He might have stated it more briefly, and more generally, as "No phenomena have properties corresponding to the concept of absolute rest."

The reason electromagnetism and mechanics get into Einstein's formulation is that the principle of relativity was already a well-known feature of mechanics. It was first enunciated by Galileo, three centuries earlier, and was built into the classical mechanics of Newton. In 1905, however, there was considerable confusion over whether the principle was applicable to electromagnetic phenomena. This accounts for the peculiar title of Einstein's paper, and his emphasis that the principle applied to both mechanical and electromagnetic phenomena. I would guess that he did not explicitly insist that the principle of relativity applied to all phenomena because in 1905 it was still possible to believe that mechanics (and gravity, often viewed at that time as a part of mechanics) and electromagnetism encompassed all the phenomena of nature. Today we know that there are other phenomena (mentioned in chapter 13), but we believe that the principle of relativity applies to all of them.

In this chapter we shall elaborate Einstein's concise statement of the principle of relativity, and then explore how the principle can be used to discover some elementary but not entirely obvious facts about how things behave. A really careful statement of the principle raises some quite subtle

---

[1] A. Einstein, "Zur Elektrodynamik bewegter Körper," *Annalen der Physik* 17 (1905): 891–921.

conceptual issues, which we will note but scrupulously avoid examining in any depth. Such a philosophical study can be entertaining, but it is distracting and of no importance for establishing a working understanding of relativity.

What *is* important is to acquire a sense of how to *use* the principle as a practical tool for enlarging one's understanding of the behavior of moving objects. Using the principle of relativity in such a way may at first be a little unfamiliar, but learning how to do it is quite unrelated to the physical and philosophical subtleties stirred up by an effort to acquire a "deep" understanding of the principle. If one wishes to understand the spectacular and counterintuitive consequences of the straightforward applications of the principle in Einstein's theory of relativity, it is essential to learn first how to apply it to some simpler, less surprising cases.

The principle of relativity is an example of an invariance principle. There are several such principles. They all begin with the phrase "All other things being the same." Then they go on to say:

1. it doesn't matter where you are. (Principle of translational invariance in space)
2. it doesn't matter when you are. (Principle of translational invariance in time)
3. it doesn't matter how you are oriented. (Principle of rotational invariance)

The principle of relativity fits into the same pattern: *All other things being the same,*

4. it doesn't matter how fast you're going if you're moving with fixed speed along a straight line. (Principle of relativity)

"It doesn't matter" means "the rules for the description of natural phenomena are the same." For example the rule describing Newton's force of gravity between two chunks of matter is the same whether they are in this galaxy or another (translational invariance in space). It is also the same today as it was a million years ago (translational invariance in time). The law does not work differently depending on whether one chunk is east or north of the other one (rotational invariance). Nor does the law have to be changed depending on whether you measure the force between the two chunks in a railroad station, or do the same experiment with the two chunks on a uniformly moving train (principle of relativity).

"All other things being the same" raises deep questions. In the case of translational invariance, it means that when you move the experiment to a new place or time you have to move everything relevant; in the case of rotational invariance you have to turn everything relevant. In the case of the principle of relativity, you have to set everything relevant into motion.

If everything relevant turned out to be the entire universe, you might wonder whether there was any content to the principle.

One can thus descend immediately into a deep philosophical abyss from which some never emerge. We shall not do this. We are interested in how such principles work on the practical level, where they are usually unproblematic. You easily can state a small number of relevant things that have to be the same and that is quite enough. When the principle doesn't work, invariably you discover that you have overlooked something else simple that is also relevant. Not only does that fix things up, but often you learn something new about nature that proves useful in many entirely different contexts. If, for example, the stillness of the air was important for the experiment you did in the railroad station, then you had better be sure that when you do the experiment on a uniformly moving train that you do not do it on an open flatcar, where there is a wind, so all other relevant things are not the same. You must do it in an enclosed car with the windows shut. If you hadn't realized that the stillness of the air was important in the station, then the apparent failure of the experiment to work the same way on the open flatcar would teach you that it was.

Invariance principles are useful because they permit us to extend our knowledge to new situations. It is on that quite practical level that we shall be interested in the principle of relativity. It tells us that no experiments that we do can enable us to distinguish between our being in a state of rest or a state of uniform motion. Any set of experiments we perform in a laboratory we choose to regard as being stationary must give exactly the same results as a corresponding set of experiments performed in a laboratory moving uniformly with respect to the first one. The results we get in the new situation, doing experiments in the uniformly moving laboratory, can be inferred from the results we found in the old situation, doing experiments in the stationary laboratory.

It is important both to understand what the principle asserts and to acquire some skill in using it to extend knowledge from one situation to another. But on a deeper level, one can again get bogged down in subtle questions. What do we mean by rest or by uniform motion? We will again take a practical view. Uniform motion means moving with a fixed speed in a fixed direction. More compactly, we say moving with a fixed *velocity*. The term "velocity" embraces both speed and direction of motion. Two boats moving 15 feet per second (f/sec), one going north and the other east, have the same speed but different velocities. I digress to remark that the foot (plural "feet") is a unit of distance (abbreviated "f"), still used in backward nations, equal to 30.48 centimeters. In this book it will be highly convenient to redefine the "foot" to be just a little shorter than the conventional English foot: about 30 centimeters (or, more precisely, 29.9792458 centimeters—98.36 percent of a conventional foot). The reasons for this redefinition will emerge in chapter 3.

It is useful to adopt the convention that a *negative* velocity in a given direction means exactly the same thing as the corresponding positive velocity in the opposite direction: −10 f/sec east is exactly the same as 10 f/sec west. Note also that in the definition of uniform motion, a fixed direction is just as important as a fixed speed: something moving with fixed speed on a circular path is not moving uniformly.

A state of nonuniform motion can easily be distinguished from a state of rest or uniform motion. You can clearly tell the difference between being in a plane moving at uniform velocity and being in a plane moving in turbulent air; between being in a car moving at uniform velocity and in one that is accelerating or cutting a sharp curve or on a bumpy road or screeching to a halt. But you cannot tell the difference (without looking out the window) between being on a plane flying smoothly through the air at 600 f/sec and being on a plane that is stationary on the ground.

In working with the principle of relativity, one uses the term *frame of reference*. A frame of reference (often simply called a "frame") is the system in terms of which you have chosen to describe things. For example, a flight attendant walks toward the front of the airplane at 3 f/sec in the frame of reference of the airplane. You start at the rear of the plane and want to catch up with him so you walk at 6 f/sec in the frame of the plane. If the plane is going at 700 f/sec, then in the frame of reference of the ground this would be described by saying that the cabin attendant was moving forward at 703 f/sec, and you caught up by increasing your speed from 700 to 706 f/sec. One of the many remarkable things about relativity is how much one can learn from considerations of this apparently banal variety.

Another important term is *inertial frame of reference*. "Inertial" means stationary or uniformly moving. A rotating frame of reference is not inertial, nor is one that oscillates back and forth. We will almost always be interested only in inertial frames of reference and will omit the term inertial except when we wish to contrast uniformly moving frames of reference to frames that move nonuniformly.

How do you know that a frame of reference is inertial? This is just another way of posing the deep question of how you know motion is uniform. It would appear that you have to be given at least one inertial frame of reference to begin with, since otherwise you can ask "Moving uniformly with respect to what?" Thus if we know that the frame in which a railroad station stands still is an inertial frame, then the frame of any train moving uniformly through the station is also an inertial frame. But how do we know that the frame of reference of the station is inertial?

Fortunately, there is a simple physical test for whether a frame is inertial. In an inertial frame, stationary objects on which no forces act remain stationary. It is this failure of a stationary object (you) to remain stationary (you are thrown about in your seat) that lets you know when the plane or car you are riding in (and the frame of reference it defines) is moving uniformly and when it is not. In our cheerfully pragmatic spirit, we will set aside the deep question of how you can know that no forces act. We will be content to stick with our intuitive sense of when the motion of an airplane (train, car) is or is not capable of making us seasick.

When specifying a frame of reference you can sometimes fall into the following trap: suppose you have a ball that (in the frame of reference you are using) is stationary before 12 noon, moves to the right at 3 f/sec between 12 p.m. and 1 p.m., and to the left at 4 f/sec after 1 p.m. By "the frame of reference of X" (also called the *proper frame* of X), one means the frame in which X is stationary. Now there is no *inertial* frame of reference in which the ball is stationary throughout its whole history. If you want to identify an inertial frame of reference as "the frame of reference of the ball," you must be sure to specify whether you mean the inertial frame in which the ball was stationary before 12, or between 12 and 1, or after 1. Depending on the time, three different inertial frames can serve as the frame of reference of the ball. Similarly for the Cannonball Express, which defines one inertial frame of reference as it zooms along a straight track at 150 f/sec from New York to Chicago, and quite another as it zooms along the same track at the same speed on the way back. The frame of reference of an airplane buffeted by high winds may never be inertial. Nor is the frame of reference of the Cannonball as it moves with fixed speed along a curved stretch of track.

Here is another, more subtle trap that many people (including, I suspect, some physicists) fall into: people sometimes take the principle of relativity to mean, loosely speaking, that the behavior of a uniformly moving object should not depend on how fast it is moving, or, to put it slightly differently, that motion with uniform velocity cannot affect any properties of an object. This is simply wrong. The principle of relativity only requires that if an object has certain properties in a frame of reference in which the object is stationary, then if the same object moves uniformly, it will have the same properties *in a frame of reference that moves uniformly with it.* But the properties that an object can have in a fixed frame of reference can certainly depend on the speed with which it moves uniformly in that frame. To take a silly example, when the object moves it has a nonzero speed, but when it is stationary its speed is zero. You could, of course, object that speed is not a property inherent in an object, but specifies a relation between the object and the frame of reference in which it has that speed. This is fine. But the nature of the trap is then that many properties that might appear to be inherent in an object turn out, on closer examination, to be relational. We shall see many examples of this.

A less trivial example is provided by the *Doppler effect.* If a yellow light moves away from you at an enormous speed, the color you see changes from yellow to red; if it moves toward you at an enormous speed, the color changes from yellow to blue. So the color of an object in a fixed frame of reference can depend on whether it is moving or at rest, and in what direction it is moving. What the principle of relativity does guarantee is that if a light is seen to be yellow when it is stationary, then when it moves with uniform velocity it will still be seen as yellow *by somebody who moves with that same velocity.*

We shall be almost exclusively interested in some simple practical applications of the principle of relativity. To apply the principle, it is essential to acquire the ability to visualize how events look when viewed from different inertial frames of reference. A useful mental device for doing this is to examine how a single set of events would be described by various people moving past them, in trains moving uniformly with different speeds.

We will be applying the principle of relativity to learn some quite extraordinary things by examining the same sets of events in different frames of reference. Some of the things we shall learn in this way are so surprising that they are hard to believe at first. You are more likely to conclude that you must have made a mistake in applying the principle. So it is quite essential to begin by acquiring some skill in using the principle of relativity to learn some things that you might not have known before, which, though not obvious, are also not astonishing. The general procedure for doing this is always the same: *Take a situation which you don't fully understand. Find a new frame of reference in which you do understand it. Examine it in that new frame of reference. Then translate your understanding in the new frame back into the language of the old one.*

Here is a very simple example. Newton's first law of motion states that in the absence of an external force a uniformly moving body continues to move uniformly. This law follows from the principle of relativity and a very much simpler law. The simpler law merely states that in the absence of an external force a stationary body continues to remain stationary.

To see how the more general law is a consequence of the simpler one, suppose we only know the simpler law. The principle of relativity tells us that it must be true in all inertial frames of reference. If we want to learn about the subsequent behavior of a ball initially moving at 50 f/sec in the absence of an external force, all we have to do is find an inertial frame of reference in which we can apply the simpler law. The frame we need is clearly the one that moves at 50 f/sec in the same direction as the ball, since in that frame of reference the ball is stationary. To put it more concretely, think of how the ball looks from a train moving at 50 f/sec alongside it. In the frame of reference of the train, the ball is stationary and we can apply the law that in the absence of an external force a stationary body remains stationary. But anything that is stationary in the train frame moves at 50 f/sec in the frame of reference in which we originally posed the problem. We conclude that since the ball remains stationary in the train frame in the absence of an external force, in the original frame it must continue to move at 50 f/sec in the absence of an external force.

So starting with the fact that undisturbed stationary objects remain stationary, we have used the principle of relativity to establish the much more general fact that undisturbed uniformly moving objects continue to move with their original velocity. (At the risk of complicating something simple, I feel obliged to remark that in reaching this conclusion we have implicitly assumed that if an object is undisturbed in one inertial frame of reference, then it is undisturbed in any other inertial frame of reference— that the condition of no force acting on an object is an *invariant* condition independent of the frame of reference in which the object is described. Since such forces can be associated with jet engines being on or off, springs being compressed or slack, etc., this is a reasonable assumption.)

If you already knew Newton's first law of motion, you might not be impressed at the power of this line of thought, so let's examine a case where what we learn might not be quite so familiar. Suppose we have two identical perfectly elastic balls. Identical elastic balls have the property that if you shoot them directly at each other with the same speed, then after they collide each bounces back in the direction it came from with the same speed that it had before the collision. Question: What happens if one of the balls is at rest and you shoot the other one directly at it?

There is a long tradition of answering such questions by invoking the conservation of energy and momentum. If you happen to know how to use such conservation laws, you should forget this for now—we shall return to the use of conservation laws in collisions in chapter 11. At this stage it is both entertaining and instructive to understand how this and many related questions can be answered using nothing but the principle of relativity. In learning how to use the principle in this way you will acquire a conceptual skill that will be essential in understanding everything that is to follow. Indeed, answering such questions using the principle of relativity provides a deeper insight than answering them by applying conservation laws.

To figure out what happens, using only the principle of relativity, first draw a picture illustrating the rule you know: when the balls move at each other with equal speeds, they simply rebound with the same speeds. This is shown in the upper part of figure 1.1. Then draw a picture of the new situation, shown in the lower part of figure 1.1. For concreteness I've taken the original speed of the moving ball to be 10 f/sec. We want to know what goes in the box in figure 1.1 with the question mark in it.

To understand what happens in the unknown case, consider it to be taking place along the tracks in a railroad station. The white ball moves to

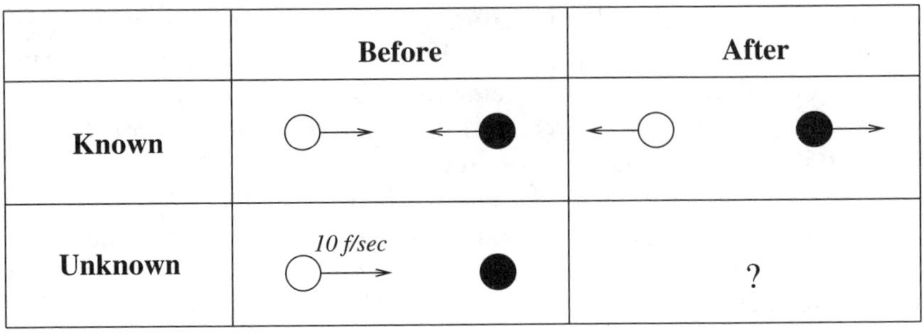

	Before	**After**
Known	◯→ ←●	←◯ ●→
Unknown	*10 f/sec* ◯→ ●	?

Figure 1.1

the right along the tracks at 10 f/sec toward the stationary black ball. This is shown again in the upper part of figure 1.2. Now think about how this would look if we were describing it from the frame of reference of the train moving through the station to the right at 5 f/sec, shown in the middle part of figure 1.2. Since the white ball covers 10 feet of track per second, and the train covers 5 feet of track per second, every second the white ball gains 5 feet on the train. So in the frame of reference of a train moving to the right at 5 f/sec, the white ball is moving to the right at 5 f/sec. Since the black ball is stationary with respect to the tracks, in the train frame it moves to the *left* at 5 f/sec, just as the tracks do. Therefore in the frame of this particular train the unknown situation before the collision, pictured just above the picture of the train in figure 1.2, becomes an instance of the known situation, pictured just below the picture of the train, in which the balls approach each other with the same speed. But the principle of relativity assures us that any experiment we do with the two elastic balls must have the same outcome in any inertial frame of reference. Since the two balls are moving at each other with the same speed—5 f/sec—in the train frame, after the collision they must bounce away from each other, each still moving at 5 f/sec in the train frame. This is also pictured just below the picture of the train.

Now all that remains is to translate this train-frame answer back to the original frame of reference—the station frame. After the collision the white ball moves to the left at 5 f/sec in the train frame, so it must be stationary in the station frame. After the collision the black ball moves to the right at 5 f/sec in the train frame, so it must be moving to the right at 10 f/sec in the station frame. This is pictured in the lowest part of figure 1.2.

So we have used the principle of relativity to learn something new about identical elastic balls: if one is at rest and the other bumps it head-on, then the moving one comes to a complete stop and the stationary one moves off with the velocity the formerly moving one originally had. This is a

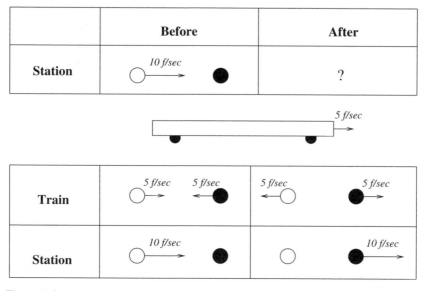

Figure 1.2

fact familiar to all players of billiards, but not many of them realize that it is simply a consequence of the much more obvious fact (less frequently encountered in billiards) that when two balls collide head-on with equal and opposite speeds, each bounces back the way it came with its original speed.

This is a dramatic illustration of the power of the principle of relativity. There is nothing very interesting about watching two balls approaching each other with equal speeds and then bouncing back, each with the same speed. If the balls are identical and sufficiently elastic, what else would you expect them to do? On the other hand, when you see a moving ball hit a stationary ball and immediately come to a complete stop, while the formerly stationary ball goes zooming off at the same speed as the originally moving ball, this can have a spectacular feel to it. One can't help wondering how the moving ball manages to come to a perfect stop and how the stationary ball manages to acquire *exactly* the same speed as the one that hit it. This mystery is solved, when you realize that the spectacular collision is just the boring one, viewed from an appropriately moving frame of reference. If you can train your imagination to experience this connection at some visceral level, then you will have mastered the principle of relativity.

Another example is shown in figure 1.3. Two identical sticky balls have the property that if they are fired directly at each other with equal speeds, then they stick together upon collision and the resulting compound ball is

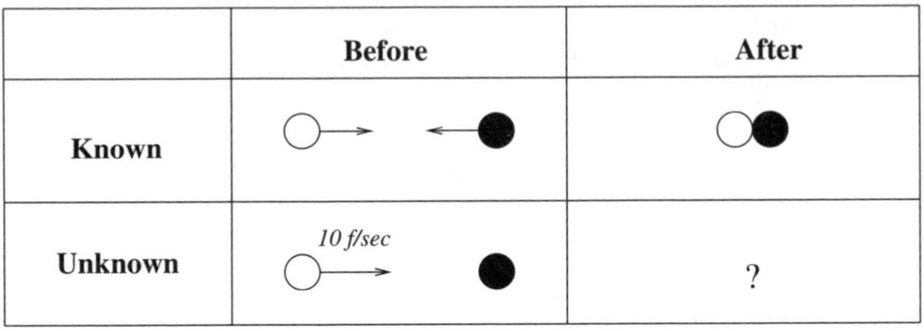

	Before	**After**
Known	◯→ ←●	◯●
Unknown	*10 f/sec* ◯⟶ ●	?

Figure 1.3

stationary. If a sticky ball is fired at 10 f/sec directly at another identical sticky ball that is stationary and the two stick together, with what speed and in what direction will the compound ball move after the collision?

We can again answer the question using only the principle of relativity, by viewing the initially moving white ball and initially stationary black ball from the frame of reference of a train in which both are moving with the same speed but in opposite directions, as shown in figure 1.4. As in our earlier example, such a train moves along the direction of motion of the white ball but only at 5 f/sec. In the train frame the situation before the collision is the one we know about: the balls move toward each other at the same speeds. Therefore in the train frame we know that after the collision the compound ball is stationary. But since the train moves down the tracks at 5 f/sec and the compound ball is stationary in the train frame, in the track frame it will move down the tracks at 5 f/sec—the same speed as the train moves in the track frame. This solves the problem: when the moving ball strikes the stationary ball, the resulting compound ball moves at half the original speed of the moving ball.

A third example is given in figure 1.5. This one has the virtue that it will not be obvious how to solve it if you happen to know about conservation of momentum, but it is easily solved using the principle of relativity. Suppose we have two elastic balls, but one of them is very big and the other is very small. If the big ball is stationary and the small ball is fired directly at it, the small ball simply bounces back in the direction it came from with the same speed, and the big ball stays at rest. (Think of throwing a table-tennis ball directly at a bowling ball.) With what speed will each ball move after the collision, if the small ball is stationary and the big ball is fired directly at it with a speed of 10 f/sec?

We wish to examine the initial situation in a frame of reference in which the big ball is stationary, so we must now view the collision in the frame of a train moving, with the big ball, at 10 f/sec to the left, as shown in figure 1.6. In that frame the small ball will move at 10 f/sec to the right,

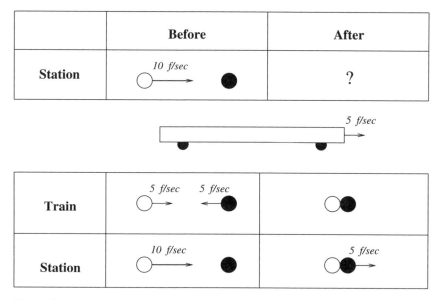

Figure 1.4

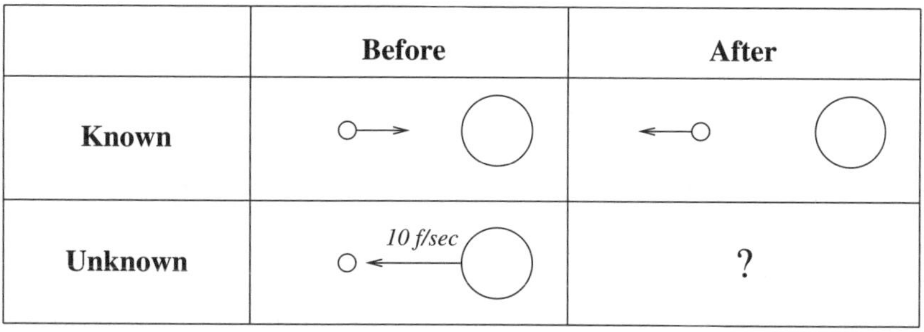

	Before	After
Known		
Unknown		?

Figure 1.5

and the situation before the collision is the one we understand. So in the train frame we know that after the collision the big ball will remain stationary and the small ball will move at 10 f/sec to the left. Returning to the description in the station frame, we note that after the collision the big ball moves with the train, at 10 f/sec to the left. The little ball, however, moves at 20 f/sec to the left, since in each second it gains 10 feet on the train, which has itself moved 10 feet to the left. So if the little ball is initially stationary, then after a collision with the big ball it moves off at *twice* the speed of the big one.

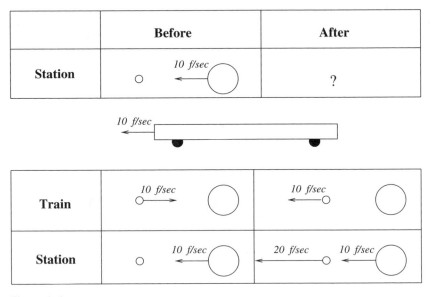

Figure 1.6

There is another interesting case to examine here, shown in figure 1.7. What happens if the big and little ball approach each other with the *same* speed—say 5 f/sec. In that case the train providing the frame of reference in which we know the answer moves, with the big ball, to the left at 5 f/sec, so the little ball moves to the right at 10 f/sec in the train frame. After the collision the big ball remains stationary in the train frame, while the little ball moves to the left at 10 f/sec. So back in the station frame the little ball moves to the left at 15 f/sec, with *triple* its original speed.

I note in passing that you can see a spectacular demonstration of this by placing the little ball, for example a tennis ball, at the very top of the big ball, for example a basketball, and then dropping them on a hard surface very carefully, so that the little ball does not roll off the top of the big one. When the big ball hits the floor it reverses its direction of motion without a change in speed, so for a very brief moment the big ball is moving up and the little ball is moving down, both going at the same speed. Immediately after that, the little ball flies up at nearly three times its original speed. As it happens, the height reached by a ball moving up is proportional to the *square* of its initial speed, so if losses due to various kinds of friction are unimportant, the little ball can shoot up to almost nine times the height from which it was originally dropped!

Seeing this has the wonderful feel of watching a good magic trick. And like a really good magic trick, when you figure out why it works—

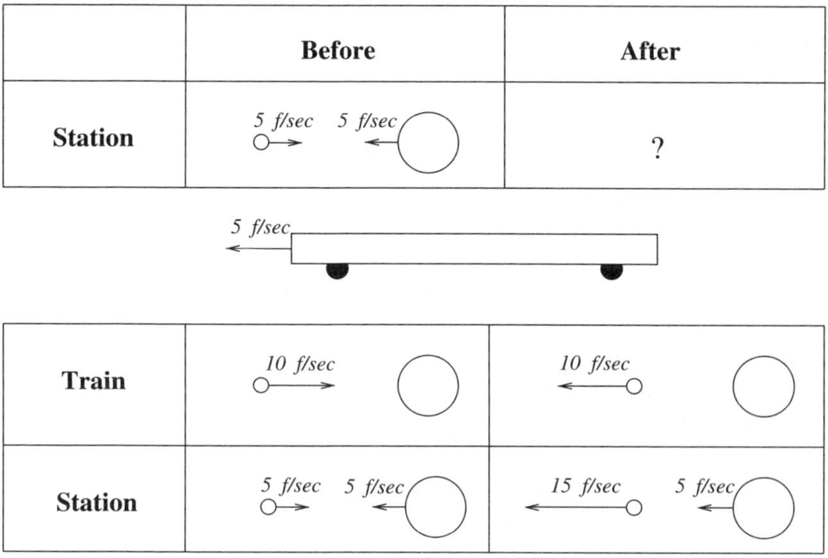

	Before	After
Station	*5 f/sec* *5 f/sec*	**?**

5 f/sec

Train	*10 f/sec*	*10 f/sec*
Station	*5 f/sec* *5 f/sec*	*15 f/sec* *5 f/sec*

Figure 1.7

figure 1.7 gives the explanation—your appreciation for the trick is only enhanced by its simplicity.

I hope these examples will give you a feeling for how the principle of relativity is actually used, and for the power it can have to predict behavior under apparently unfamiliar conditions. Before starting to apply it under genuinely unfamiliar conditions, we must look a little more closely at some of the reasoning we used in these simpler examples.

Two

Combining (Small) Velocities

IN CHAPTER 1 WE EXAMINED the power of the principle of relativity, deducing the not entirely obvious outcomes of certain collisions by considering other collisions whose outcomes were self evident. But I must now emphasize that besides using the principle of relativity, we repeatedly made implicit use of another rule that enabled us to relate the velocity of a ball in the train frame to its velocity in the station frame. This short chapter gives an explicit statement of this rule.

If, as I hope is the case, the rule strikes you as obvious and therefore uninteresting, please bear with me. My reason for belaboring the boringly obvious is that in this case what appears to be obvious is not, in fact, exactly correct. The rule is accurate to a phenomenally high degree of precision when all relevant velocities are no more than many thousands of feet per second, but when velocities become as large as many millions of feet per second, we shall find, quite surprisingly, that the rule has to be modified.

The rule we implicitly made use of in chapter 1 goes under the name of the *nonrelativistic velocity addition law*. "Nonrelativistic" is an unfortunate term, but everybody uses it and so shall we. It does not mean, as you might think, "in contradiction to the principle of relativity." It comes from the fact that the body of lore constructed by applying the principle of relativity to certain strange facts about motion at very high speeds has come to be known as the theory of relativity. As a result, the term "nonrelativistic" refers to how we thought the world behaved before we learned about the theory of relativity. Since at low speeds things actually do behave almost exactly the way we used to think they did before we learned about the theory of relativity, "nonrelativistic" means valid to a high degree of accuracy when all speeds are sufficiently small. How small will emerge in subsequent chapters, but note in this regard that even the speeds of bullets from guns (many thousands of feet per second) count as very small indeed.

Before stating the nonrelativistic velocity addition law, we must establish a convention on the direction of motion along which velocities are taken to be positive. For almost all of the points we shall be making, it suffices to consider objects confined to move along a single direction, which we shall often take to be that of a long straight railroad track. There are only two possible directions of motion along such a track, and we need names for them. Let us therefore take the tracks to run east-west and agree that motion to the east is assigned a positive velocity, while motion to the west is given a negative velocity. Thus a ball going west at a *speed* of 5 f/sec has, according to our convention, a *velocity* of −5 f/sec.

There is a certain amount of subtlety (but not very much) in these matters. A velocity is always defined with respect to a frame of reference. Let a train move east at 10 f/sec. Suppose a ball moves toward the rear of the train at 3 f/sec so that in the track (or station) frame it moves east at only 7 f/sec. In the track frame the velocity of the ball is +7 f/sec, since it is moving east. But in the train frame the ball is moving toward the rear of the train—i.e. toward the west. So its train-frame velocity is −3 f/sec. It is important to keep in mind that in this context "toward the west" means "toward the western end of the train." West is a direction, not a place. If the train is heading from California to New York, the ball is getting farther from California (which many people think of as "the West") even though it is moving toward the western end of the train. This, of course, is because the train is moving away from California at a higher speed than the ball is moving west in the train frame.

In pictures of events along the track in various frames of reference, we shall take the tracks to be more or less horizontal and shall follow the convention of mapmakers, taking east to be to the right, and west to be to the left.

Another possibly pedantic remark should be made, before we codify the "obvious" way we transformed velocities from the train frame to the station frame and back again. Let X be an object (a ball, if you like) moving uniformly along the tracks. There is a frame of reference in which the velocity of X is 0—in which X is stationary. This frame is, of course, the frame of reference of a train that is moving uniformly along the tracks with the same velocity as X. As noted in chapter 1, such a frame is called the proper frame of X. The term "proper" should not be construed as meaning that there is anything particularly virtuous about this particular frame of reference. It's just that every uniformly moving object does have a unique frame associated with it in a natural way —namely, the frame in which it does not move. (Rest is always to be regarded as a special case of motion: motion with zero speed.) Often one leaves out the word "proper" and just refers, for example, to the frame of the ball or the ball frame. If an object is moving nonuniformly then there is no *inertial* frame in which it remains stationary at all times; a frame of reference in which a nonuniformly moving object remains stationary must be a noninertial frame.

If Y is a second object moving uniformly along the tracks with a velocity different from the velocity of X, then we can, if we wish, describe the motion of Y in the proper frame of X, calling Y's velocity in X's frame v_{YX}. The expression v_{YX} is read as "the velocity of Y with respect to X." It doesn't matter whether we think of either Y or X as being an object or a frame of reference—the proper frame of the associated object—since both the object and the associated frame of reference move with the same velocity.

With this, we can now state in all its abstract glory the nonrelativistic velocity addition law. If X, Y, and Z all move with uniform velocity along the same straight line, then

$$v_{XZ} = v_{XY} + v_{YZ}. \tag{2.1}$$

In words, the velocity of X with respect to Z is the sum of the velocity of X with respect to Y and the velocity of Y with respect to Z. Or, if you prefer, the velocity of X in frame Z is the sum of the velocity of X in frame Y and the velocity of frame Y in frame Z.

Suppose, for example, X is a ball, Y is a train, and Z is the station (tracks). Then (2.1) says that the velocity of the ball in the station frame is the velocity of the ball in the train frame plus the velocity of the train in the station frame. This should be evident when all the velocities are positive. If the ball moves 5 f/sec east in the train frame and the train moves 10 f/sec east in the station frame, then the ball moves 15 f/sec (5 f/sec + 10 f/sec) to the east in the station frame. But it also works when some of the velocities are negative. If instead the ball moves toward the rear of the train at a speed of 3 f/sec, so that its velocity v_{XY} in the train frame is -3 f/sec, then its velocity in the station frame is 7 f/sec (-3 f/sec + 10 f/sec). If the ball moves toward the rear of the train at a speed of 16 f/sec, then its velocity in the station frame is -6 f/sec (-16 f/sec + 10 f/sec). Now, even in the station frame, it is moving west at a speed of 6 f/sec.

Usually, of course, it's simpler just to reason one's way to the answer without having to invoke the abstract form (2.1) of the addition law. It looks, irritatingly, like one of those formulas that have been written down merely for the sake of having a formula: a needless petty display of erudition. Soon, however, we shall learn that the addition law (2.1) is not exactly correct, being valid to an extremely high degree of accuracy when all speeds are not too large. When enormous velocities enter the story, or if we want the right answer to fantastically high precision, then we must use a modified form of (2.1), and it is then important to use the formula that expresses the modified addition law, since commonsense reasoning no longer gives the right answer. So you should be sure you understand how to use the nonrelativistic addition law (2.1), even when what it tells you is "obvious."

One important consequence of (2.1) (which turns out to remain valid at any speed) is that

$$v_{XY} = -v_{YX}. \tag{2.2}$$

If X moves with a certain speed with respect to Y, then Y moves with that same speed with respect to X, but in the opposite direction. This (fairly obvious) relation follows directly from the general rule (2.1). For consider the special case in which Z and X are identical, so that v_{XZ} becomes v_{XX}, the velocity of X in the frame in which X is stationary, i.e. in its proper frame. The velocity of X in its proper frame is 0, so we have

$$0 = v_{XX} = v_{XY} + v_{YX}, \tag{2.3}$$

and this immediately gives (2.2).

How would one go about justifying the rule (2.1) to a stubborn person who did not find it obvious? Consider this instance of it: Let a train move east in the track frame. If a ball moves east in the train frame at 5 f/sec, then in one second the ball gets 5 feet nearer the front of the train. And if the train moves at 10 f/sec in the track frame, then in one second the train gets 10 feet further east along the track. So in one second the ball gets 15 feet further east along the track—the 5 it gains on the train and the additional 10 the train gains on the track. But the ball getting 15 feet further east along the track in one second is precisely what we mean when we say the ball moves at 15 f/sec in the track frame. Who could doubt this? Indeed, I encourage you not to doubt it until you find boringly familiar the points made in this and the preceeding chapter.

But I do call your attention to an apparently innocent phrase that turns out, surprisingly, to be fraught with danger: *in one second*. We have implicitly assumed that "in one second" means the same thing in the train frame as it does in the track frame. "Well," you will say, "of course it does. A second's a second. My time is your time." But suppose that were not true. Suppose "in one second" in the train frame meant something different from "in one second" in the track frame. What would happen to the argument we just gave? We would have to replace "in one second" by something like "in one second according to train time" or "in one second according to track time." The argument we just went through then starts off fine, but is a bit more cumbersome:

> If the ball moves east in the train frame at 5 f/sec then in one second according to train time it gets 5 feet further down the train. And if the train moves at 10 f/sec in the track frame, then in one second according to track time it gets 10 feet further east along the track.

But then we come to:

> So *in one second* the ball gets 15 feet further east along the track—the
> 5 it gains on the train and the additional 10 the train gains on the track.

What can that italicized "in one second" mean here? The first 5 feet are
gained in one second according to train time, the second 10 feet are gained
in one second according to track time. Collapsing both into a single,
unqualified "in one second" makes no sense unless track time and train
time are the same.

For the moment, we will not pursue this any further. But be aware that
the simple rule (2.1) telling us how velocities combine relies on the implicit
assumption that there is nothing problematic about the idea of a single
unique notion of time that can be used equally well in any frame of reference. It was Einstein's great insight in 1905 that this apparently obvious
assumption is, in fact, false. "It came to me," he said to a colleague many
years later, "that time was suspect." When the assumption of a unique
frame-independent time fails, it takes other "obvious" assumptions down
with it.

That failure, however, is so slight as to be of no importance when all
speeds of interest are small compared with that of light, as they were in
the examples we examined in chapter 1. So now we must turn to how the
speed of light enters the story.

Three

The Speed of Light

WHEN YOU TURN ON A LIGHT, how long does it take the light to get from the bulb to the things it illuminates? Galileo apparently tried to answer this by stationing two people with lanterns on top of two mountains, a large distance D apart. Alice opens her lantern, Bob opens his the instant he sees Alice's, and Alice notes the time T that passes between the moment she opens hers and the moment she sees the light returning to her from Bob's. To get the speed c with which the light moves from her mountaintop to Bob's and back again, Alice just divides twice the distance between the mountains by the delay time T to get

$$c = 2D/T. \tag{3.1}$$

I don't know if Galileo worried about it, but there is a problem here: only part of the delay is due to the time it took the light to get from Alice to Bob and back. The rest is due to the speed of Bob's response—the time it takes the reception of Alice's light at Bob's eyes to reach his brain and be converted into a signal that reaches the muscles in his arms that operate the tendons that cause his fingers to open the shutter of his lantern.

There is an easy and clever way to take care of this problem. Do the experiment again with Bob on a third mountain farther away from Alice. Bob's response time won't change if the light from Alice is not so much dimmer that it is significantly harder to see, so the increase in the delay time is entirely because of the extra time it takes the light to travel the extra distance to and back from the more distant mountain. Since this extra time is just twice the extra distance divided by the speed of light, Alice is now able to figure out the speed of light without having to know anything about Bob's response time. She simply uses (3.1) with D being the increase in the distance between her and Bob in the two cases, and T being the increase in the time between her sending and receiving light signals.

But unfortunately, if she does this, Alice will observe no discernible change in the delay time. Either it takes no time at all for light to travel the extra distance—i.e., the speed of light is infinite—or Bob's sluggish response takes so very much longer than the light travel time that Alice simply can't tell the difference between the two cases. And indeed, light travels so quickly and human response times are, by comparison, so sluggish, that two terrestrial mountains within view of each other are much too nearby for this method to work.

Three centuries later, Galileo's unsuccessful attempt was realized by replacing the two mountains by the Earth and the Moon. The Moon is so far away that it takes radar more than 2 seconds to get there and bounce back. But by then the speed of light was known to high precision by other methods. Note, by the way, that the speed of radar is the same as the speed of light. Both are forms of electromagnetic radiation and all forms of electromagnetic radiation (light, radar, radio, x-rays, gamma rays, TV signals, for example) have the same speed in empty space.

Light travels so fast that to measure its speed either you have to let it travel an enormous distance, or you have to make very accurate measurements of extremely tiny intervals of time. The very first successful estimate of the speed of light came from using astronomical distances. Galileo, who plays many roles in the story of relativity, discovered the four major moons of Jupiter earlier in the 17th century. In 1676 careful observations by Ole Rømer, of the regularly occurring moments when a moon disappeared within Jupiter's shadow, revealed that sometimes these Jovian lunar eclipses lagged behind schedule by about 10 minutes, and sometimes they came in 10 minutes ahead. It was noted that they were ahead of schedule when the Earth was closest to Jupiter and behind when the Earth was furthest away. One concludes that the time it takes light to cross the orbit of the Earth must be something like 20 minutes. This gives an estimate of several hundred thousand kilometers per second for the speed of light.

In 1849 Hippolyte Louis Fizeau performed a terrestrial measurement by sharpening up the precision with which tiny time intervals could be measured. Imagine an axle with identical cog wheels at each end, perfectly aligned so that a thin beam of light moving parallel to the axle through a gap between the teeth of one wheel is able to get through the corresponding gap in the other wheel. If you now spin the whole thing extremely rapidly about the axle, you might hope that during the (very tiny) time it takes the light to pass between the two wheels, the second wheel will have turned enough for the gap in the second wheel to have moved out of alignment enough to block at least some of the light that passed through the gap in the first wheel. After all, the wheels are spinning extremely fast and the teeth of the second wheel have to move only a tiny fraction of a full turn to start to block some of the light.

It turns out that for an axle short enough not to disturb, by a little twisting or bending, the rather delicate alignment of the two cog wheels, the light still travels much too fast for this to work for any feasible rate of spinning. But one can also introduce an enormous time-consuming detour for the light, in the form of a periscope-like perpendicular side journey with the help of four mirrors. When this was done (the detour was several kilometers in Fizeau's experiment), the sought-for effect was observed, and the resulting estimate for the speed of light was in good agreement with that furnished by the earlier astronomical measurement of Rømer.

Today we have highly sophisticated ways to measure the speed of light and know that it is 299,792,458 meters per second (m/sec). Furthermore, that is what it always shall be, because as of 1983 the meter has been *defined* to be not the distance between two scratches on a platinum-iridium bar lovingly cared for in Paris, but as the distance light travels in 1/299,792,458 of a second. Our unit of length (the meter) now is tied to our unit of time (the second). You might think that since the speed of light is now fixed forever by definition of the meter, this means that there is no longer any point in striving to measure it more and more accurately. But such improved experiments now provide more and more accurate measurements of the length of a meter—better and better standards of length. The experiments remain just as important as they used to be. What has changed is how we describe what we have learned from them.

There are two useful numerical near coincidences associated with the speed of light being 299,792,458 m/sec:

First, the number is extremely close to 300 million m/sec or 300,000 kilometers per second (km/sec). Physicists are very used to taking it to be 3×10^8 m/sec. So much so that there is a legend that somebody once fouled up the report of a fine high-precision experiment by using the number 3 rather than 2.998 in converting the result into a more convenient form.

Second, the corresponding English unit is about 186,000 miles per second. Since there are 5,280 feet in a mile, there is good news for those in Washington and a few remote outposts elsewhere in the world who still resist the metric system. For this works out to about 982,000,000 feet per second. Thus within 2 percent accuracy the speed of light is 1 billion feet per second or, in more practical units, 1 foot per nanosecond (f/ns). (A nanosecond is a billionth of a second.) A speed of 1 f/ns is actually relevant in setting limits to the size a computer can have if you want it to be really fast. Arithmetic operations are now being done in substantially less than a microsecond (a millionth of a second), nanosecond computers are just around the corner, and if you want to exchange information with another part of the computer before you do the next operation, it had better not be more than half a foot away, since (as we shall see) no information can be transmitted faster than the speed of light. Nanoseconds and feet are also relevant to the accuracy of the global positioning system (GPS), which uses satellites broadcasting time signals every nanosecond. The signals are therefore spaced a foot apart as they arrive at the surface of the Earth, establishing the foot as a measure of the accuracy of the system.

In thinking about relativity it is very convenient to measure speeds in units that assign an especially simple value to the speed of light. In 1959 the foot was officially defined to be exactly 0.3048 of a meter. Since the speed of light is exactly 299,792,458 m/sec, if only people in 1959 had defined the foot to be 0.299792458 of a meter, a mere 1.64 percent shorter, then the speed of light would now be *exactly* 1 f/ns. This unit of length will prove to be so useful, that for the purposes of this book *I hereby redefine the foot:*

Henceforth, by 1 foot we shall mean the distance light travels in a nanosecond. A foot, if you will, is a light nanosecond (and a nanosecond, even more nicely, can be viewed as a light foot). We can revert to the clumsier term light nanosecond if it is ever necessary to distinguish between our foot, and the conventional slightly larger foot. If it offends you to redefine the foot (as it did one referee of a paper I sent to the *American Journal of Physics*), then you may define 0.299792458 meters to be 1 phoot, and think "phoot" (conveniently evocative of the Greek $\phi\omega\tau o\varsigma$, "light") whenever you read "foot."

For comparison with lesser speeds, it can sometimes be convenient to think of 1 foot per nanosecond as 1,000 feet per microsecond. Since the speed of sound in ordinary air is about 1,000 f/sec and a microsecond is a millionth of a second, this shows that light travels about a million times faster than sound.

There is something peculiar and, when you think about it, quite extraordinary about the unqualified assertion that the speed of light in empty space is 299,792,458 m/sec. Ordinarily when you specify a speed to such high precision and indeed when you mention any speed at all, the question "with respect to what" comes irresistibly to mind. After all, the speed of an object depends on the frame of reference in which that speed is measured. As we have repeatedly noted, a ball that Alice throws while riding on a uniformly moving train has one speed with respect to the train but quite another speed with respect to the tracks. In the case of light there are two obvious possible answers to the question "with respect to what?":

First Obvious Answer

The speed of light is 299,792,458 m/sec with respect to the source of the light. When you turn on a flashlight, the light it produces has a speed of 299,792,458 m/sec with respect to that flashlight. What else could it be? In much the same way, when one specifies the speed of a bullet, one always has in mind its speed with respect to the gun from which it has emerged.

This reasonable answer is contradicted by our current understanding of the electromagnetic character of light. In the 19th century there was a great unification of the laws of electricity and magnetism, completed by the Scottish physicist James Clerk Maxwell. Maxwell's equations led to the prediction that when electrically charged particles jiggle back and forth (as they do, for example, inside a hot wire) they must emit radiant energy that travels at a speed of about 300,000,000 m/sec. Since this speed was numerically indistinguishable from the speed of light, it was natural to identify light with a particular form of such radiation (associated with a very rapid jiggling—almost a million billion times a second). Maxwell's equations imply quite unambiguously that this speed does not depend on the speed of the source of the radiation. According to the theory, the speed of the light is the same whether the chunk of matter in which the charged particles are jiggling is stationary or moving toward or away from the direction in which the light is emitted.

People had also noted that the regularity of certain astronomical motions as observed on Earth was quite unaffected by whether the source of the light that enabled us to observe them was moving toward or away from us, as it would be if the light traveled to us more slowly when the source was moving away than when the source was moving toward us. So there was both theoretical and astronomical evidence that the speed of light did not depend on the speed of its source.

Second Obvious Answer

With respect to a light medium (historically called the ether), I emphasize that 299,792,458 m/sec is the speed of light *in vacuum*. Light goes significantly slower in transparent media like water or glass, and a little bit slower in air. This ether, then, would be a sort of irreducible residue of otherwise empty space—what remains after you've removed everything it is possible to remove.

The analogy now is not to bullets from a gun, but to sound, which is a wave in the air. Like the speed of light, the speed of sound does not depend on the speed of the source of the sound. Sound moves at a definite speed with respect to the air, whose vibrations constitute and transmit that sound. If light is a vibration of something called the ether, then the speed of light should be with respect to that ether.

Since the Earth moves about the sun at a brisk clip of 30 km/sec in different directions, depending on the time of year, and the sun moves briskly about the center of our galaxy, it would be a remarkable coincidence if the Earth just happened to be stationary in the rest frame of the ether. One would expect there to be a kind of "ether wind" blowing past the Earth, leading to a dependence of the speed of light on Earth on the direction of that wind. The speed of light on Earth into the direction from which the ether wind was blowing ought to be a bit less than its speed along the direction of the wind. Efforts to detect such a difference failed to yield a clear-cut result, most famously in the Michelson-Morley experiment of 1887. The measurements demonstrated that if the speed of light was fixed with respect to an ether, then the Earth, in spite of its complicated motion with respect to the galaxy, was improbably close to being at rest in the rest frame of that ether at the time the experiment was performed. Stubborn people considered the possibility that the Earth dragged the ether in its neighborhood along with it. But if that were so, then the apparent positions of stars in the sky should shift through the year depending on the way in which the ether was being dragged by the Earth. No such shift was observed.

The importance of the Michelson-Morley experiment in the historical development of relativity has been debated. Einstein apparently alludes to it in his famous 1905 paper setting forth relativity, but only once and then only in passing: "Examples of this sort, together with *unsuccessful attempts to determine any motion of the earth relative to the 'light medium,'* lead to the conjecture that . . . " (my italics). The reference is little more than parenthetical. Such attempts had to be mentioned, because had they been successful and unambiguously demonstrated a significant direction dependence to the velocity of light on Earth, the theory of relativity would have been dead on arrival.

The "examples of this sort" that Einstein offered as the real motivation for his reexamination of the nature of time were all examples of the fact that the electric and magnetic behavior of matter is consistent with the principle of relativity, in spite of the then widespread view that there actually was a preferred inertial frame of reference for electromagnetic phenomena—the frame in which the ether was stationary. The equations of electromagnetic theory were held by many to be valid in that frame of reference and no other. Einstein noted, in effect, that even if this were so, a broad range of electromagnetic phenomena seemed to play out in exactly the same way in frames of reference other than the frame in which the ether was stationary. This led him to postulate that the laws of electromagnetism were, in fact, rigorously valid in arbitrary inertial frames of reference. If this postulate were valid then, Einstein noted, "the introduction of a 'luminiferous ether' will prove to be superfluous" because there would be no way to determine the rest frame of the ether by any physical experiment involving electromagnetic phenomena. It is this specific postulate—that what we now call the principle of relativity applies to electromagnetism as well as to Newtonian mechanics (where everybody agreed that it was indeed valid)—that Einstein named the "principle of relativity" (*Prinzip der Relativität*).

Now if Maxwell's equations are valid in any inertial frame of reference, and if they predict that electromagnetic radiation and light in particular propagate at a fixed speed that is independent of the speed of the source of the light, then light must propagate at the same speed in any inertial frame of reference. The answer to the question "with respect to what?" is, as we now know, "with respect to any inertial frame you like." The speed of light in vacuum is simply 299,792,458 m/sec in any inertial frame of reference, regardless of how fast the source of the light is moving, and regardless of the choice of frame of reference in which the measurement of the speed of the light is made. If, for example, you race after the light in a rocket at 10 km/sec, you do not reduce its speed away from you to 299,782 km/sec. It still recedes from you at 299,792 km/sec. I emphasize that it is only the speed of light in *vacuum* that has this special property. The speed of light in water *does* depend on how fast you are moving with respect to the water, though not in an obvious way, as we shall see. Indeed, what is special here is not light, but the speed $c = 299,792,458$ m/sec. When one says "speed of light" without any qualification, one almost always means the speed of light in vacuum, 299,792,458 m/sec.

How can this be? How can there be a speed c with the property that if something moves with speed c then it must have the speed c in any inertial frame of reference? This fact—known as the *constancy of the speed of light*—is highly counterintuitive. Indeed, "counterintuitive" is too weak a word. It seems downright impossible. One of the central aims of this book is to remove this sense of impossibility and to see how it can, in fact, make perfect sense.

A nomenclatural digression: Today everybody calls the speed of light in vacuum c, as in, most famously, "$E = mc^2$," about which there will be more to say in chapter 11. I once thought c stood for "constant," reflecting the fact that it doesn't vary from one frame of reference to another. But perhaps it stands for *celeritas*—Latin for "speed"—as in "celerity" or "accelerate."

To make sense of the constancy of the speed of light we must look very closely and critically at what it actually means to "have a speed" with respect to a particular frame of reference. When we say that an object moves uniformly with a certain speed s, we mean that it goes a certain distance D in a certain time T and that the distance and time are related by $D/T = s$. We are thus led to examine carefully how one actually measures such distances and how one actually measures such times.

Let P be a valid procedure for carrying out the time and distance measurements that allow one to determine the speed of an object in a given inertial frame. Let Bob, carrying out the procedure P in the frame of reference of a space station, measure the speed of a pulse of light as it zooms off into space. He will find that it moves at about 299,792 km/sec. Suppose Alice flies swiftly after the light at a speed Bob determines to be 792 km/sec. Bob will then (correctly) note that in each second the light gets an additional 299,792 km away from him and Alice gets an additional 792 km away, so that the distance between Alice and the light is growing at only 299,000 km/sec. But if Alice carries out the same procedure P in the frame of reference of her rocket ship, she will find that the speed of the light is 299,792 km/sec, so that in her own frame of reference the distance between her and the light is still growing at the full 299,792 km/sec.

How are we to account for this discrepancy? Obviously the methods Alice uses to measure distances and times must be different from those used by Bob. But don't they both use exactly the same procedure P? Yes, but you have to think about what "exactly the same" means. If Bob, for example, uses clocks that are stationary in the frame of his space station to measure times, then if Alice uses exactly the same procedure in her frame of reference, she must use clocks that are stationary in the frame of *her* rocket ship. Thus in Bob's frame of reference Alice's clocks are moving, while his are not, and, of course, vice versa: in Alice's frame Bob's clocks are moving and hers are not. Similar considerations apply to the meter sticks they might use to measure distances. The not terribly subtle but easily overlooked point is that Bob's procedure *as described in Bob's frame of reference* must be exactly the same as Alice's procedure *as described in Alice's frame of reference.* But Alice's procedure as described in *Bob's* frame of reference is not exactly the same as Bob's procedure as described in *Bob's* frame of reference.

It is this difference that makes it possible for either Bob or Alice to account, in an entirely rational way, for the discrepancy in their conclusions. The fact that Alice and Bob, using different frames of reference, both find exactly the same speed for one and the same pulse of light appears paradoxical only if you make several assumptions about the relation between the clocks and meter sticks used by Alice and Bob. Before 1905 everybody implicitly made all of these assumptions:

1. The procedure Alice uses to synchronize all the clocks in her frame of reference gives a set of clocks that Bob agrees are synchronized when he tests them against a set of clocks that he has synchronized using the same procedure in his own frame of reference. ("Same" here, as earlier, is to be taken to mean that what Bob does has the same description in his own frame of reference as Alice's procedure has in hers.)
2. The rate of a clock, as determined in Bob's frame of reference, is independent of how fast that clock moves with respect to Bob.
3. The length of a meter stick, as determined in Bob's frame of reference, is independent of how fast that meter stick moves with respect to Bob.

If any of these assumptions is false, then we must reexamine the nonrelativistic velocity addition law—the rule specifying how the speed of an object changes as one changes the frame of reference in which its speed is measured. Today we know that *all three* of these assumptions are false. The special theory of relativity gives a quantitative specification of how they fail, and how, when they are suitably corrected, one emerges with a simple and coherent picture of space and time measurements that is entirely in accord with the existence of an invariant speed—a speed that is the same in all inertial frames of reference.

The traditional (and simplest) way to arrive at this picture—the way we shall be taking and the way Einstein used—is simply to accept as a working hypothesis that in any inertial frame of reference, any procedure that correctly measures the speed of light in vacuum *must* give 299,792,458 m/sec. We shall accept the strange fact that if Alice and Bob both measure the speed of one and the same pulse of light, they will both find it to be 299,792,458 m/sec, even though Alice and her measuring instruments may be moving in the same direction as the light with respect to Bob and his. By tentatively accepting this peculiar fact, and insisting that the principle of relativity must remain valid, we will be able to *deduce* the precise way in which each of the three assumptions about the behavior of moving clocks and meter sticks must be modified. Once this is done, and the corrected versions of these three assumptions are identified and understood, the strange fact will cease to appear strange. More importantly, we will have acquired a firm understanding of the new and wonderful subtleties Einstein first realized about the nature of time.

This remarkable property of light—that its speed does not depend on the frame of reference in which it is measured—is today called the *principle of the constancy of the velocity of light*. The special theory of relativity is said to rest on two principles: the principle of relativity and the principle of the constancy of the velocity of light. In Einstein's great 1905 paper, he did not use the word *Prinzip* for this second principle (as he did for the first). He characterized each principle as a "postulate" (*Voraussetzung*). His second postulate was that light in empty space moves with a velocity that is independent of the velocity of the body that emitted the light. This is tantamount to the second principle when it is conjoined with the first, which Einstein stated as the postulate that the concept of absolute rest has no more meaning for electromagnetic phenomena than it does for phenomena in ordinary Newtonian mechanics.

If you know a little German and are interested in seeing how Einstein sets up the problem, you can download the text of his famous paper in which relativity first appears, "Zur Electrodynamik bewegter Körper," cited at the beginning of chapter 1, through a link that can be found at `http://press.princeton.edu/titles/8112.html` (with everything I've quoted being on the first page and a half).

Four

Combining (Any) Velocities

IN CHAPTER 2 WE ARGUED that if Alice, a passenger on a train moving at v feet per second, can throw a ball at u feet per second, then if she throws the ball toward the front of the train, its speed w with respect to the tracks will be

$$w = u + v \tag{4.1}$$

in the same direction as the train.

This is known as the nonrelativistic velocity addition law. It is called "nonrelativistic" because it is only accurate when the speeds u and v are small compared to the speed of light. Evidently it fails to work when $u = c$ (i.e. if Alice turns on a flashlight instead of throwing a ball) for we know that the speed w of the light in the track frame will not be $c + v$ but simply c—the same value it has in the train frame!

Suppose, however, that Alice fired a gun that expelled bullets whose muzzle velocity u was 90 percent of the speed of light. The "bullets," if you insist on getting practical about it, could be pulses of light, traveling down the train in a pipe containing a fluid in which the speed of light was only 0.9 feet per nanosecond. (It is only the speed of light *in vacuum* that is the same in all frames of reference.) If the addition law (4.1) fails when $u = c$, it would be surprising if the law worked very well when u was $0.9c$—and in fact it does not. Both (4.1) and the frame independence of the special velocity c, turn out to be special cases of a very general rule for compounding velocities that works whether or not the speeds involved are small compared to the speed of light. This *relativistic velocity addition law* states that

$$w = \frac{u + v}{1 + \left(\frac{u}{c}\right)\left(\frac{v}{c}\right)}. \tag{4.2}$$

If u and v are both small compared with the speed of light, then u/c and v/c are both small numbers. Their product is then a small fraction of a small number—i.e. a *very* small number—so the relativistic rule (4.2) differs from the more familiar nonrelativistic rule (4.1) only by dividing the nonrelativistic result by a number that differs insignficantly from 1. If, on the other hand, $u = c$, then (4.2) requires w also to be c, whatever the value of v may be. (Check this for yourself! It's an easy algebraic exercise.) Thus (4.2) is consistent with our nonrelativistic experience—i.e. our experience in dealing with situations in which all relevant speeds are small compared to the speed of light—as well as with the speed of one and the same pulse of light being the same in all inertial frames of reference.

We shall now show that the more general relativistic rule (4.2) is a direct and immediate consequence of the constancy of the velocity of light and the principle of relativity. We shall find that if the speed of light is the same in all inertial frames of reference, then the addition law (4.1) *must* be replaced by (4.2) regardless of what kind of moving objects we are describing and regardless of how fast they are moving. That so much more general a rule follows from the special case of the constancy of the velocity of light, together with the principle of relativity, is a remarkable demonstration of the power of that principle. In its scope, the argument that will lead us to (4.2) is analogous to our extraction of Newton's first law of motion in chapter 1 by applying the principle of relativity to the fact that stationary bodies remain stationary in the absence of an applied force. But while it is obvious, once you have understood the principle of relativity, that the special case of stationary objects remaining stationary implies the general rule that uniformly moving objects continue in the same state of uniform motion, the connection between the special case of the constancy of the speed of light and the much more general rule (4.2) is far from obvious.

Before we embark on this important application of the principle of relativity, you might note that the explicit occurrence of the speed c in (4.2), even when none of the objects or frames of reference associated with u, v, or w have anything to do with light, gives an early indication that the speed c is built into the very nature of space and time. Things that move at that special speed move at that speed in all frames of reference, as a direct consequence of (4.2) itself. Pulses of light in vacuum happen to be examples of such things. But the speed c has an importance that goes beyond the fact that light moves at that speed in empty space.

To develop a strategy for deducing the relativistic addition rule (4.2), we must first ask what goes wrong when we try to justify the nonrelativistic rule (4.1). The obvious way to determine the speed of an object is to determine the time it takes it to traverse a racetrack of known length. Doing this requires two clocks, placed at the two ends of the racetrack, to determine the exact times at which the object starts and finishes the race. To arrive at the nonrelativistic velocity addition law (4.1), we implicitly assumed that people using the train frame and people using the track frame would agree on whether those two clocks were synchronized. Prior to Einstein this essential assumption was never explicitly noted. Although people realized that it could be difficult as a practical matter to arrange for two clocks in faraway places to be synchronized, they took it for granted that there was nothing about it that was problematic in principle.[1] We also implicitly assumed that the people using different frames of reference would agree on the length of the racetrack between the two clocks and on the rates at which the clocks were running.

The constancy of the velocity of light means that the nonrelativistic addition law (4.1) cannot be correct for an object moving at the speed of light, and therefore it means that at least some of the assumptions on which (4.1) rests must be wrong. This, in turn, casts doubt on the validity of the nonrelativistic addition law for any velocities at all. But if we are not allowed to make such assumptions about the basic instruments with which we measure velocities, how can we deduce the correct rule for compounding velocities? One way to arrive at it would be to figure out, and then take fully into account, a set of new "relativistic" rules about clock-synchronization disagreements, rates of moving clocks, and lengths of moving measuring sticks, but this takes a bit of doing. It is, in fact, the usual way of arriving at the correct relativistic addition law (4.2) in most expositions of the subject. Although we will eventually construct the new set of rules about clocks and measuring sticks, at this stage we don't know any of them. Nevertheless, it is possible and useful to figure out the correct velocity addition law before learning anything about the behavior of moving clocks and measuring sticks, and this is the path we shall follow.

The direct way to get at (4.2) is to take advantage of the fact that we do know the speed of at least one thing: light. By being clever we can use light to help us measure the speed of anything else in a way that makes no use whatever of either clocks or measuring sticks. This enables us to deduce the rule for how velocities change when the frame of reference changes, without assuming anything at all about their behavior. The idea is to let the moving object—call it a ball—run a race with a pulse of light—call it a photon. By comparing how far the ball goes with how far the photon goes, we can figure out the speed of the ball. If, for example, the photon, moving at speed c, covers twice as much ground as the ball, then the speed of the ball must be $\frac{1}{2}c$. (We shall consider only the case in which the ball

---

[1]For a survey of the practical difficulties and importance of the synchronization of distant clocks at the time Einstein put forth relativity, see Peter Galison, *Einstein's Clocks, Poincare's Maps: Empires of Time* (New York: W. W. Norton) 2003.

goes slower than the photon. Later we will see that there is something highly problematic about balls that move faster than light.)

This neat idea runs into an immediate difficulty. Although the photon and the ball start their race in the same place, they will be in different places at the end of the race. But to compare how much ground they cover during the race, we must be able to determine exactly where the ball is at the precise moment the photon reaches the finish line. To do this we need two synchronized clocks, one at the finish line and one with the ball. We can then determine where the ball is at the moment the photon reaches the finish line, by noting where the ball is when its clock reads exactly the same time that the clock at the finish line reads at the moment the photon gets to the finish line. But this requires knowing whether two clocks in different places are synchronized—precisely the issue we wished to avoid.

There is an easy way around this problem. Rather than end the race when the photon reaches the finish line, we arrange for it to hit a mirror and bounce back the way it came. We end the race only when the photon reencounters the ball, which is still moving in its original direction. By ending the race when the photon and the ball arrive at the same place, we solve the problem of determining, without clocks, just where the ball is along its path when the race ends. At the moment the race ends, the ball is precisely where the photon is.

Suppose this is all done on a train. We first describe the race using the train frame. Let the race start at the rear of the train and let the photon be reflected back toward the rear when it reaches the front. Suppose the photon meets the ball a fraction f of the way from the front of the train back to the rear. (If, for example, the train consists of 100 identical cars, numbered $1, 2, 3, \ldots$ starting from the front, and the photon meets the ball in the passageway between cars 34 and 35, then $f = 0.34$.) Between the beginning and the end of the race, the photon has gone the entire length of the train *plus* an additional fraction f of that length, but between the beginning and end of the race, the ball has only gone the entire length of the train *minus* that same fraction f of the length. The ratio of the distance covered by the ball to the distance covered by the photon is thus $\frac{1-f}{1+f}$. This is pictured in figure 4.1.

Since the photon and the ball were both racing for the same time, this ratio must also be the ratio of their speeds. I emphasize that the fact that they were both racing for the same time is now unproblematic, requiring no clocks to establish it, because we have organized the race so that the photon and the ball are in the same place when they start the race and also when they finish it. So if we call the velocity of the ball in the train

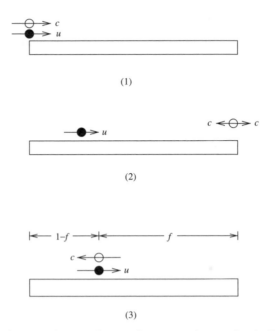

Figure 4.1. A photon (white circle, speed c) runs a race with a ball (black circle, speed u) in a stationary train (long rectangle). The race is pictured at three different moments. (1) At the start of the race the photon and the ball are together at the rear of the train, moving with speeds c and u. (2) The photon reaches the front of the train and bounces back toward the rear (whence the two-headed arrow). (3) At the conclusion of the race the photon reencounters the ball a fraction f of the way back from the front of the train.

frame u, then, since the speed of the photon in either direction is c,

$$\frac{u}{c} = \frac{1-f}{1+f}. \tag{4.3}$$

The people on the train have thus measured the speed of the ball without using clocks and without having to know the length of the cars in their train. They only have to be able to count cars. If the ball met the photon some fraction of the way along a car, they would have to be able to compare the lengths of the two parts of the car, but they could do this without knowing the absolute length of either part by just counting up the number of times some measuring stick of unknown length went into both parts. So (4.3) summarizes a simple way to compare the velocities of two objects, which avoids using any clocks and avoids having to know any absolute distances. It is useful to rewrite (4.3) as a relation that expresses the fraction f in terms of the speed u of the ball and the speed of light c:

$$f = \frac{c - u}{c + u}. \qquad (4.4)$$

[A pedagogical digression: When I assert that two expressions are equivalent—in this case, the relation (4.3) and the relation (4.4)—you should convince yourself that I'm telling the truth. If the algebraic exercise of deriving (4.4) from (4.3) proves tiresome, you should at least convince yourself in a few special cases. If, for example, $f = \frac{1}{2}$ then (4.3) tells us that $u/c = \frac{1}{3}$. On the other hand, when $u = \frac{1}{3}c$, (4.4) does indeed give $f = \frac{1}{2}$.]

Now let us start all over again and analyze a similar race on the train, but this time using the frame of reference of the track, in which the train has a velocity v (which we take to be less than the speed of light) and the ball has a velocity w. We take u, v, and w all to be positive—i.e., the ball moves to the right in the train frame, and the train and ball move to the right in the track frame—so that velocities and speeds are the same; the result we shall arrive at, however, turns out to be valid for any combination of positive and negative velocities. As before, the photon and ball both start at the rear of the train, the photon reaches the front first and bounces back toward the rear, and the race ends when the photon reencounters the ball. We again want to know what fraction of the way back along the train the photon has to go before it meets the ball. We want to express this fraction entirely in terms of the various speeds. This time the analysis is a bit more complicated, since the train is moving during the race.

We continue to assume that the photon moves with speed c in both directions in the track frame. In a little while we are going to appeal to the constancy of the velocity of light to interpret this as exactly the same race as the one we analyzed in the train frame. Meanwhile, however, it might be a good idea to put the first race out of your mind while analyzing this one. You may think, if you want, of the photon in the second race as a new "track-frame photon" that has the speed c in the track frame, unlike the old train-frame photon, which had the speed c in the train frame. If you look at it this way (and you should for now) then there is nothing at all peculiar about the track-frame analysis that follows. It's more elaborate than the train-frame analysis, but only because now the train is moving too, which complicates things.

To analyze the race in the track frame we shall have to talk about track-frame distances and times. We shall not, however, make any assumptions about how track-frame clocks and measuring sticks behave, except that track-frame people have taken all necessary precautions to ensure that the track-frame speed of an object is indeed the track-frame distance it goes in a given track-frame time. Our goal is to end up with relations like (4.3) or (4.4) that involve no times and lengths. The relation we seek involves only velocities, together with the fraction f of the way back along the train the photon has to go before it meets the ball. All of the unknown distances and times we introduce will drop out at the end.

Suppose (see figure 4.2) it takes a time T_0 for the photon to get from the back of the train to the mirror at the front and a time T_1 for the reflected photon to get from the front to the point a fraction f of the way back along the train where it reencounters the ball. Let L be the length of the train and let D be the distance between the front of the train and the ball at the moment the photon reaches the front of the train. All these times and distances are unknown track-frame times and distances, but since the reasoning that follows is entirely track-frame reasoning, and since the problematic quantities D, L, T_0, and T_1 all drop out of the final result, this causes no difficulty. These times and distances are illustrated in figure 4.2, which you should return to, when appropriate, in the course of the discussion that follows.

Since T_0 is the time it takes the photon to get a distance D ahead of the ball and since both start in the same place at the same moment and move toward the front with speeds c and w, we must have

$$D = cT_0 - wT_0. \tag{4.5}$$

On the other hand, T_1 is the time it takes the photon and ball, initially a distance D apart, to get back together. Since the photon covers a distance cT_1 during this time and the ball, wT_1, we have

$$D = cT_1 + wT_1. \tag{4.6}$$

Since we don't know the value of D, we shall eliminate it from these two relations. This gives us $cT_0 - wT_0 = cT_1 + wT_1$, which it is convenient to write in the form

$$\frac{T_1}{T_0} = \frac{c - w}{c + w}. \tag{4.7}$$

But unfortunately we don't know the times T_1 and T_0. There is, however, a second quite similar way to get at the same ratio of these two times, by comparing the progress of the photon not with that of the ball, as we have just done, but with that of the train. Note first that T_0 is the time it takes the photon to get ahead of the rear of the train by the length L of the train. Since the photon has speed c and the train, speed v,

$$L = cT_0 - vT_0. \tag{4.8}$$

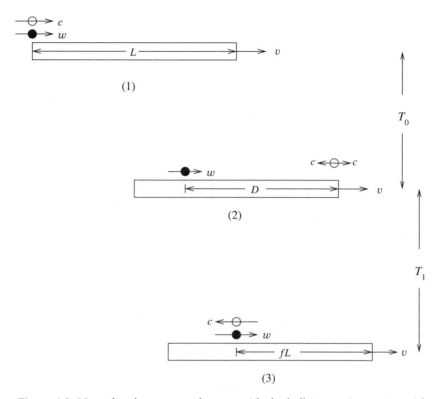

(1)

T_0

(2)

T_1

(3)

Figure 4.2. Now the photon runs the race with the ball in a train moving with speed v. The race is pictured at three different moments from the point of view of the track frame. (1) At the start of the race the photon and the ball are together at the rear of the train moving with speeds c and w. The length of the train is L. (2) A time T_0 after the events pictured in (1), the photon reaches the front of the train and bounces back toward the rear. At this moment the photon is a distance D ahead of the ball. (3) At the conclusion of the race, a time T_1 after the events pictured in (2), the photon reencounters the ball a fraction f of the full length of the train—i.e., a distance fL—back from the front of the train.

Note next that T_1 is the time it takes the photon, moving toward the rear at speed c, to meet a point on the train originally a distance fL away from it that moves toward it at velocity v. Thus

$$fL = cT_1 + vT_1. \tag{4.9}$$

We don't know the actual value of L any more than we knew the actual value of D, but we can also eliminate L from these last two equations. This gives us $cT_1 + vT_1 = f(cT_0 - vT_0)$, which gives us a second expression for the ratio of T_1 to T_0:

$$\frac{T_1}{T_0} = f\left(\frac{c-v}{c+v}\right). \tag{4.10}$$

Although we don't know either T_1 or T_0, this expression for their ratios must be the same as the other expression (4.7). We conclude that

$$f\left(\frac{c-v}{c+v}\right) = \frac{c-w}{c+w}. \tag{4.11}$$

This is the relation we need. All unknown times and distances have dropped out, and we have a relation involving only the fraction f and some velocities. It follows immediately from (4.11) that the fraction f can be expressed in terms of the velocities v and w by

$$f = \left(\frac{c+v}{c-v}\right)\left(\frac{c-w}{c+w}\right). \tag{4.12}$$

I stress that as a piece of track-frame analysis, applicable to a race between a ball with track-frame speed w and a photon with track-frame speed c, both of them on a train with track-frame speed v, there is nothing at all peculiar about the analysis leading to (4.12). As a reassuring check that we haven't made some mistake, notice the following. Suppose the velocity v of the train in the track frame were 0. Then the track frame would be the same as the train frame. Consequently w, the velocity of the ball in the track frame, would be the same as u, the velocity of the ball in the train frame. And, indeed, if you set v to zero and take w to be u, you do get back our old train-frame result (4.4). Galileo would have been quite happy with the argument leading to (4.12) (though we would have to turn the train into a boat.) Indeed, the result (4.12) remains entirely correct if we replace the photon by anything at all that moves with the same speed in both directions, and allow that speed, c, to be any speed at all greater than w and v.

We do something that would not have been to Galileo's liking only when we now declare that if the photon really *is* a photon, and if the speed c really is the speed of light in vacuum, then these two pieces of analysis we have now completed can be taken to be train-frame and track-frame analyses of *one and the same race*. In this race u is the train-frame velocity of the ball, w is the track-frame velocity of that same ball, and v is the track-frame velocity of the train. Peculiarly, however—and this is the *only* peculiarity in the entire argument—we now insist that the track-frame speed c of that one photon (in either direction) is exactly the same as the train-frame speed c of that same photon (in either direction). In both frames (and in both directions) that speed is c—1 foot per nanosecond. This is the only place in the entire argument where we invoke the counterintuitive principle of the constancy of the velocity of light.

But if we have indeed been describing one and the same race in two different frames, then f, the fraction of the way back from the front of the train where the photon meets the ball, must have the same value in either frame. For although there might (and indeed, as we shall see, there will) be disagreement between the two frames of reference over the length of the cars of the train, there can be no disagreement about where on the train the photon meets the ball. Their reunion could trigger an explosion, for example, that would make a smudge on the floor, which all observers in all frames could inspect later on at their leisure to confirm in which part of which car the meeting took place.

So the track-frame expression (4.12) for the fraction f must agree with the train-frame expression (4.4). Setting them equal gives us a relation between the three velocities w, u, and v:

$$\left(\frac{c+v}{c-v}\right)\left(\frac{c-w}{c+w}\right) = \frac{c-u}{c+u}. \qquad (4.13)$$

It is useful to rewrite this relativistic velocity addition law in a form, like the form of the nonrelativistic velocity addition law (4.1), in which w appears on the left side and u and v on the right:

$$\frac{c-w}{c+w} = \left(\frac{c-u}{c+u}\right)\left(\frac{c-v}{c+v}\right). \qquad (4.14)$$

This is the relativistic rule that replaces the nonrelativistic rule (4.1). Instead of *adding* u and v to get w we must *multiply* an expression involving u by an expression of the same form involving v to get a third expression of the same form involving w.

The relation between the nonrelativistic rule (4.1) and the relativistic rule (4.14) is not at all clear. To see that they are, in fact, rather simply related, one must carry out the elementary algebraic exercise of solving (4.14) for the velocity w of the ball in the track frame in terms of its speed u in the train frame and the speed v of the train. The result is the relativistic velocity addition law stated in (4.2):

$$w = \frac{u+v}{1 + \left(\frac{u}{c}\right)\left(\frac{v}{c}\right)}. \qquad (4.15)$$

If you find it too complicated to carry out the algebra taking you from (4.14) to (4.15), you can find it done at the end of this chapter.

Although the two forms (4.14) and (4.15) of the velocity addition law are entirely equivalent ways of expressing the same relation among the three velocities w, u, and v, it is helpful to keep them both in mind, since one form can be more useful than the other, depending on the question one is asking. Thus the form (4.15) makes immediately evident (as noted at the beginning of this chapter) why the nonrelativistic addition law $w = u + v$ becomes quite accurate when u and v are small compared to the speed of light. The form (4.14), on the other hand, directly reveals the following important fact:

If the speed u of the ball in the train frame and the speed v of the train in the track frame are both less than the speed of light, then both $\frac{c-u}{c+u}$ and $\frac{c-v}{c+v}$ will be numbers between 0 and 1. Since the product of two numbers between 0 and 1 is also between 0 and 1, this means that $\frac{c-w}{c+w}$ is between 0 and 1, which implies in turn that the speed w of the ball in the track frame is also less than the speed of light. This conclusion also follows from (4.15), of course, but it is more evident as a consequence of (4.14).

Thus the obvious stratagem for producing an object moving faster than light does not work: if you have a cannon that shoots balls at 90 percent of the speed of light, and you put it on a train moving at 90 percent of the speed of light, pointing toward the front of the train, then the speed of the ball in the track frame will still be less than the speed of light. Indeed, in this particular case (4.15) tells us that the speed w of the ball in the track frame will be a fraction $\frac{0.9+0.9}{1+(0.9)^2} = \frac{1.80}{1.81}$ of the speed of light—about 99.45 percent. This is the first indication we have found—there will be several others—that no material object can travel faster than the speed of light.

For many purposes it is helpful to abstract the relativistic addition law from the context of balls, trains, and tracks and state it in terms of the velocities of certain objects (or frames of reference) with respect to other objects (or frames of reference). Let us regard the track as an object called A, the train as an object called B, and the ball as an object called C. The velocity v of the train in the track frame we now call v_{BA}—the velocity of B with respect to A. In the same way we call the velocity u of the ball in the train frame v_{CB}, the velocity of C with respect to B, and we call the velocity w of the ball in the track frame, v_{CA}. In this language the two forms for the addition law become

$$\frac{c - v_{CA}}{c + v_{CA}} = \left(\frac{c - v_{CB}}{c + v_{CB}}\right)\left(\frac{c - v_{BA}}{c + v_{BA}}\right) \tag{4.16}$$

and

$$v_{CA} = \frac{v_{CB} + v_{BA}}{1 + \left(\frac{v_{CB}}{c}\right)\left(\frac{v_{BA}}{c}\right)}. \tag{4.17}$$

Another advantage of (4.16) over (4.17) emerges when you consider the case in which object C is a rocket that itself emits a fourth object D. If D has speed v_{DC} with respect to C what is the speed v_{DA} of D with respect to A? In other words, what form does the addition law take when we compound three speeds instead of just two? This leads to a great mess if we try to answer the question using the form (4.17), but if we use the addition law in the form (4.16), we merely note the following:

The speed of D with respect to A can be arrived at by compounding the speed of D with respect to C and the speed of C with respect to A. Applying the general rule (4.14) to this case gives

$$\frac{c - v_{DA}}{c + v_{DA}} = \left(\frac{c - v_{DC}}{c + v_{DC}}\right)\left(\frac{c - v_{CA}}{c + v_{CA}}\right). \tag{4.18}$$

But now we can apply (4.14) again to express the quantity containing v_{CA} in terms of v_{CB} and v_{BA} to get

$$\frac{c - v_{DA}}{c + v_{DA}} = \left(\frac{c - v_{DC}}{c + v_{DC}}\right)\left(\frac{c - v_{CB}}{c + v_{CB}}\right)\left(\frac{c - v_{BA}}{c + v_{BA}}\right). \tag{4.19}$$

So to compound three speeds rather than just two, we just put a third term into the product in (4.16) to get (4.19). Evidently if D were a rocket that emitted a fifth object E, we could continue in this way, and so on indefinitely. The rule in the form (4.17) would get more and more complicated, but in the form (4.16) it would retain the same simple form.

The addition law in either of its two forms (4.17) or (4.16) continues to hold even when not all the velocities have the same sign—even, for example, when the ball moves toward the rear of the train rather than the front. If Alice throws a ball with speed u toward the *rear* of a train that moves with positive velocity v along the track, then the velocity w of the ball along the track is given by

$$w = \frac{-u + v}{1 - \frac{u}{c}\frac{v}{c}} \tag{4.20}$$

since this is what (4.2) reduces to when u is replaced by $-u$. (It is a useful exercise to check this by repeating the analysis of this chapter for the case where the race starts at the front of the train rather than at the rear.)

An easier if more abstract way to see that (4.16) and (4.17) remain valid even when all the velocities are not positive is to note that, although we have derived (4.16) in the case where all three of the velocities v_{CA}, v_{CB}, and v_{BA} are positive, we can introduce negative velocities by exploiting the general fact that

$$v_{YX} = -v_{XY}. \qquad (4.21)$$

We can, for example, express the positive v_{BA} in (4.16) in terms of the negative velocity v_{AB}, writing $v_{BA} = -v_{AB}$. Having done this we can then algebraically transform (4.16) into

$$\frac{c - v_{CB}}{c + v_{CB}} = \left(\frac{c - v_{CA}}{c + v_{CA}}\right)\left(\frac{c - v_{AB}}{c + v_{AB}}\right). \qquad (4.22)$$

Notice that this has exactly the same form as (4.16)—only the labels A and B have been interchanged. But now one of the three velocities, v_{AB}, is negative.

Similar tricks using (4.21) enable one to reexpress (4.16) in other equivalent forms in which either or both of the two velocities on the right are negative.

Remarkably, an experiment confirming the validity of the relativistic velocity addition law (4.2) was performed in 1851 by Fizeau, a few years after he measured the speed of light and more than half a century before Einstein wrote his 1905 paper in which the relativistic addition law first appears. Fizeau's experiment measured the speed of light in moving water.

As we have noted, the speed of light in stationary water is less than its speed c in vacuum. Traditionally it is written as $\frac{c}{n}$ where n, called the "index of refraction of water," is about $1\frac{1}{3}$, so that the speed of light in water is about $\frac{3}{4}$ of its speed in vacuum. From the nonrelativistic addition law one would expect that if water flowed down a pipe with velocity v, then the speed of light in the moving water would be its speed in stationary water increased by the speed of the water in the pipe: $w = \frac{c}{n} + v$. What Fizeau observed in 1851, however, was

$$w = \frac{c}{n} + v(1 - \frac{1}{n^2}). \qquad (4.23)$$

Fizeau's result (4.23) was viewed as a confirmation of a rather elaborate contemporary ether-theoretic calculation based on the idea that the water was partially successful in dragging the ether along with it. But it is much more simply understood today as an elementary consequence of the relativistic velocity addition law:

If w is the speed of light in the frame of the pipe, $\frac{c}{n}$ is the speed of light in the frame of the water, and v is the speed of the water in the frame of the pipe, then (4.17) tells us that

$$w = \frac{\frac{c}{n} + v}{1 + \frac{v}{nc}}. \tag{4.24}$$

This form for w looks quite different from (4.23), but to see the resemblance let us consider not w, but the amount by which w differs from the speed of light in stationary water, $w - \frac{c}{n}$. It follows directly from (4.24) that

$$w - \frac{c}{n} = \frac{\frac{c}{n} + v - \frac{c}{n}\left(1 + \frac{v}{nc}\right)}{1 + \frac{v}{nc}} = \frac{v\left(1 - \frac{1}{n^2}\right)}{1 + \frac{v}{nc}}. \tag{4.25}$$

This is precisely Fizeau's result (4.23), except that the term $v\left(1 - \frac{1}{n^2}\right)$ giving the effect of the moving water is divided by $1 + \frac{v}{nc}$. But the speed v of the water in the pipe is, of course, a tiny fraction of the speed c of light in vacuum, so this divisor differs imperceptibly from 1. The experiment would have to be done with impossibly high precision to reveal the tiny difference between (4.23) and (4.25). Although Einstein does not mention Fizeau's experiment in his 1905 paper, he later said that it was of fundamental importance in his thinking.[2]

It remains to give the algebraic demonstration that the form (4.15) of the relativistic velocity addition law does indeed follow from the alternative form (4.14):

Write (4.14) in the form

$$\frac{c - w}{c + w} = \frac{a}{b}, \tag{4.26}$$

where

$$a = (c - u)(c - v) \tag{4.27}$$

and

$$b = (c + u)(c + v). \tag{4.28}$$

---

[2] See R. S. Shankland, "Conversations with Albert Einstein," *American Journal of Physics* **31** (1963): 47–57. It is therefore pleasing to see Fizeau's result emerge as an immediate consequence of Einstein's two postulates.

It follows from (4.26) that

$$(c - w)b = (c + w)a \tag{4.29}$$

or

$$c(b - a) = w(b + a) \tag{4.30}$$

so that

$$\frac{w}{c} = \frac{b - a}{b + a}. \tag{4.31}$$

Now according to (4.28) and (4.29)

$$\begin{aligned} b &= c^2 + c(u + v) + uv, \\ a &= c^2 - c(u + v) + uv, \end{aligned} \tag{4.32}$$

and therefore

$$\begin{aligned} b + a &= 2(c^2 + uv) = 2c^2\left(1 + \left(\tfrac{u}{c}\right)\left(\tfrac{v}{c}\right)\right), \\ b - a &= 2c(u + v). \end{aligned} \tag{4.33}$$

The two relations in (4.33) immediately reduce (4.31) to (4.15).

> In the remainder of this chapter we reexamine some of the collisions discussed in chapter 1 when speeds are not small compared with the speed of light, using the relativistic velocitiy addition law (4.2) instead of the nonrelativistic law (4.1). While I would encourage you at least to glance at these examples, they are not essential for the developments that follow.

As we noted in chapter 2, our use in chapter 1 of the principle of relativity to predict the outcomes of some simple collisions relied on the nonrelativistic velocity addition law. The conclusions we reached there can therefore be reliable only when all velocities are small compared with that of light. To predict the outcomes when the velocities are not small, we can apply the same approach, being careful, however, to replace all uses of the nonrelativistic velocity addition law with the correct relativistic law.

Consider, for example, the collision of the big and little elastic balls depicted in figures 1.5 and 1.6. In the frame in which the big ball is stationary, the little ball simply bounces back in the direction it came from with its original speed, the big ball remaining stationary. Suppose this holds even when the speed of the little ball is comparable to the speed of light c. We again ask what happens to the little ball if it is originally stationary and the big ball is fired at it with speed u. In the nonrelativistic case we found that after the collision the little ball moved with speed $2u$. If u were bigger than half the speed of light, this might appear to be a good way to get the little ball moving faster than light, but now, of course, we must reexamine the problem using the relativistic addition law.

Suppose the big ball moves to the right with speed u in the frame in which the little ball is initially stationary. In the proper frame of the big ball, the little ball moves to the left with speed u, so in that frame after the collision the little ball bounces back to the right with the same speed u. To learn how the little ball moves after the collision in the original frame, we use the relativistic addition law (4.15), taking u to be the post-collision velocity of the little ball in the frame of the big ball, and v to be the velocity of the big ball in the frame in which the little ball is initially at rest, which is also u. So what (4.15) tells us is that w, the post-collision velocity of the little ball in the frame in which it was initially at rest, is given by

$$w = \frac{u + u}{1 + \left(\frac{u}{c}\right)\left(\frac{u}{c}\right)} = \frac{2u}{1 + \left(\frac{u}{c}\right)^2}. \tag{4.34}$$

When u is a tiny fraction of the speed of light c, we have our old nonrelativistic answer: the small ball moves off with twice its original speed. But if $u = \frac{1}{2}c$, (4.34) tells us that after the collision it moves off not at speed c but at $\frac{4}{5}c$. If $u = \frac{3}{4}c$, after the collision the little ball moves off not at one and a half times the speed of light but at $\frac{\frac{3}{2}c}{1+\frac{9}{16}} = \frac{24}{25}c$. And indeed, you can easily convince yourself (from the fact that $(1 - \frac{u}{c})^2$ is positive) that $2u/c$ is always less than $1 + \left(\frac{u}{c}\right)^2$. So no matter how fast the big ball is thrown at it, the little ball always bounces away at less than the speed of light.

We can make a similar reexamination of the case in which the big and little balls are fired at each other with the same speed u. In the nonrelativistic case we found that the little ball bounced away from the big one at three times its original speed, as shown in figure 1.7. Now when we go to the rest frame of the big ball, the speed w of the little ball before the collision will be given by (4.34). In the rest frame of the big ball, the little ball bounces back in the direction it came from with the same speed w. When we use the relativistic velocity addition law to transform this back to the original frame, we find that the speed of the little ball after the collision in the original frame is

$$\frac{w + u}{1 + \left(\frac{w}{c}\right)\left(\frac{u}{c}\right)}. \tag{4.35}$$

With w given by (4.34) this becomes

$$\frac{3 + \left(\frac{u}{c}\right)^2}{1 + 3\left(\frac{u}{c}\right)^2}\, u. \tag{4.36}$$

When $\frac{u}{c}$ is very small this is indeed very close to the nonrelativistic result that the little ball bounces off with a speed of $3u$, but when u is $\frac{1}{3}c$, it is only $\frac{7}{9}c$, and even when $u = \frac{2}{3}c$, it is only $\frac{62}{63}c$.

Five

Simultaneous Events; Synchronized Clocks

THE PUZZLEMENT WE FEEL at the fact that a given pulse of light has the same speed in both the track frame and the train frame can be traced to a deeply ingrained misconception about the fundamental nature of time. Until we learn otherwise—and prior to Einstein in 1905, nobody had learned otherwise—we implicitly believe that there is an absolute meaning to the simultaneity of two events that happen in different places, independent of the frame of reference in which the events are described. This assumption is so pervasive in our view of the world that it is built into the very language we speak, making it difficult to reexamine the question of what it actually means to assert that two events in different places are simultaneous.

Before embarking on such a reexamination, it is necessary to take a careful look at the term *event*, which plays a fundamental role in the relativistic description of the world. An event is something that happens at a definite place at a definite time. It is, if you like, the space-time generalization of the purely spatial geometric notion of point.

If you look at this a little more closely, you realize that, like the concept of a point, the concept of an event is an idealization. No object we can actually get our hands on has the property of zero spatial extension that characterizes a geometric point, and no process we will be talking about has zero extension in both space and in time. (Although zero extension in time is captured by the word "instantaneous," I know of no English word—other than "pointlike," if that is a word—that signifies zero extension in space.) Whether or not we wish to view something as an event depends on the spatial and temporal differences we want to discriminate between. If, for example, the relevant time scale is years and the relevant distance scale is hundreds of kilometers, then it makes sense to view the meeting of a class between 1:25 and 2:40 in room 115 of Rockefeller Hall on the Cornell University campus in Ithaca, New York, as an event (at least in frames of reference that are not moving too rapidly with respect to the Earth). But if the relevant scales are minutes and feet, it does not.

So a phenomenon can be viewed as constituting a single event in a given frame of reference, if its temporal and spatial extension in that frame are both small compared with all other times and distances of interest. All of the events we will be examining will qualify as events in all the frames of reference we are interested in—they can be viewed as space-time points.

How can we decide whether two different events, happening in different places, that are simultaneous in the train frame are also simultaneous in the track frame? To be concrete, suppose one event consists of quickly making a tiny mark on the tracks (as they speed past) from the rear of the train, and the other consists of doing the same from the front. The two events could be anything else you like: bells being rung at the front and rear of the train, lightning bolts striking each end, etc. But since it will be useful to mark the spot along the tracks where each event occurs, it is simplest to take the two events to be nothing more than two acts of marking the tracks.

How does Alice, who uses the train frame, persuade herself that the two marks on the tracks are made at the same time? Well, she could provide both ends of the train with accurate clocks and confirm that each mark was made when the clock at its end of the train read noon. But how can she be sure that the two clocks are properly *synchronized?* How does she know that they both read noon *at the same time?*

Trying to check the simultaneity of the two events with clocks gets Alice nowhere, since confirming that the clocks are properly synchronized requires her to have precisely what we're trying to construct: a way to confirm that two events in different places—in this case, each clock reading noon—happen *at the same time.*

This is a centrally important point. It is useful to have two clocks in different places only if they are properly synchronized. But "synchronized" means that the clocks have the same reading *at the same time.* Therefore you need a way to check that two events in different places are simultaneous, if you want to check that the two clocks in different places are synchronized. The question of whether clocks in different places are synchronized, and the question of whether events in different places are simultaneous, are simply different aspects of the same fundamental puzzle. You can answer one question if and only if you can answer the other.

Let us try again. Alice could bring the two clocks together, directly confirm that they read the same when they're both in the same place, and then have them carried to the two ends of the train. But how does she know how fast each clock was running as it was carried to its end of the train? Faced with a phenomenon as peculiar as the constancy of the velocity of light, it would be rash to assume that she knew anything about the rate at which a clock runs when it is moving. (We will learn how to deal with this in chapter 6.) The straightforward way to check on whether the clocks have done anything peculiar while being carried to the two ends of the train would be to compare what each reads when it gets there with the reading of a stationary clock at that end of the train. But we can only do this if those two stationary clocks are properly synchronized, which brings us right back to the same problem.

Ah, but suppose, even though we don't know how it might affect their rates, the two clocks start at the exact middle of the train and are carried to the two ends in exactly the same way except, of course, that one of the two clock-transportation procedures is executed in the opposite direction from the other. Then, however erratically its motion caused one clock to behave during the journey, the other, having experienced just the same kind of trip, would have run erratically in exactly the same way. So even if they lost or gained time because of their motion, the two clocks would still agree when they arrived at the ends of the train. That method of providing both ends of the train with synchronized clocks ought to work. And it does! In the train frame.

But now we are faced with another problem. Even if Alice did cleverly use two such centrally synchronized, symmetrically transported clocks to confirm that two events at the two ends of the train were simultaneous in the train frame, Bob, using the track frame, would not agree that the the two clock-transportation procedures were identical, because in the track frame motion toward the front of the train is *not* insignificantly different from motion toward the rear.

Bob *would* agree with Alice that the reading of one of her clocks, the instant it arrived at its end of the train, was the same as the reading of the other clock, the instant it arrived at the other end, since Alice and Bob can't disagree about things that happen *both* in the same place *and* at the same time. But Bob need not agree with Alice that the identical readings of the clocks meant that it took an identical amount of time for each clock to get to its end, since the clocks were not moving symmetrically in the track frame and therefore might be running at different rates during their journeys from the center to the ends. So Bob will have to do a rather elaborate calculation to determine whether each clock reached its end of the train *at the same time* as the other clock. That calculation would have to figure out how fast each of the clocks was moving in the track frame, and how far it had to go. It could get quite complicated. It can, however, be done, and it leads to a remarkable conclusion that we can extract by a much simpler stratagem.

The simpler stratagem, like our earlier method for finding the relativistic velocity addition law, avoids all worries about possibly misbehaving clocks by using in the train frame a method to check the simultaneity of two events in different places that makes no use of clocks at all. This method can easily be analyzed in the track frame too. The analysis relies only on the fact that the speed of light is c—1 foot per nanosecond— regardless of the direction the light is moving in and regardless of the frame of reference in which the speed is measured.

Why, you might ask, should we build such a strange fact into our procedure for determining whether two spatially separated events are simultaneous? If you do ask, then you have forgotten why we started worrying about whether simultaneity might depend on frame of reference. It was our hope that this might lead us to a clearer understanding of the constancy of the velocity of light. What we are doing is perfectly sensible. We *start* from the strange fact of the constancy of the velocity of light, and see what it *forces* us to conclude about the simultaneity of events. We shall find that it forces us to conclude that whether two events in different places are simultaneous does indeed depend on the frame of reference in a way that can be stated simply and precisely.

Note first that Alice, on the train and using the train frame, can easily exploit the fact that light travels with a definite speed c to arrange that the two marks on the tracks are made from the two ends of the train simultaneously. She places a lamp exactly halfway along the train, and then turns on the lamp. Light from the lamp races toward both ends of the train at the same speed c. Since the light has to travel the same distance (half the length of the train) in either direction, and moves at the same speed in either direction, it arrives at the two ends of the train *at the same time*. So if the making of each mark on the tracks is triggered by the arrival of the light, the marks will certainly be made at the same time.

Alice has thus managed to produce a pair of events in different places that are simultaneous (in the train frame) without having to make any problematic use of clocks. This procedure, of course, works for any two signals that move from the center of the train to the two ends, as long as they move at the same speed. If the common speed of the signals is not the speed of light, however, the track-frame analysis cannot be as concise as it is with light signals, because the two signals have different speeds in the track frame. The two speeds can be found with the help of the relativistic velocity addition law, and with considerably more effort one can generalize to arbitrary signals the analysis based on light signals. One then finds (see pp. 54–56) that this more general way to produce two simultaneous events in the train frame leads to a relation between the track-frame time and track-frame distance between the two events that is exactly the same as the relation we now extract with much less effort using light signals.

The question facing us is how this procedure, which convinced Alice that the two events are simultaneous in the train frame, will be interpreted by Bob, who uses the track frame. Bob will certainly agree that the lamp is indeed in the center of the train, for if the train is 100 cars long and the lamp is bolted down between cars 50 and 51, then there is no denying that it is indeed in the center. (This is true even if the length of the train in the track frame is altered by its motion—as we shall see in chapter 6 it is—because whatever that alteration might be, it would be the same for both the front half and the rear half of the train.) But in the track frame when the lamp is turned on and the light starts to move toward the two ends, the rear of the train moves toward the place where the light originated and the front moves away from it. Since the speed of the light in either direction is c in the track frame—remember we are using the strange fact that the speed of the light is 1f/ns in the track frame even though it is also 1 f/ns in the train frame—in the track frame it will clearly take the light less time to reach the rear of the train, which is heading toward the light to meet it, than it will take the light to reach the front of the train, which is running away as the light pursues it.

So Bob must conclude that the light reaches the rear of the train before it reaches the front, and therefore that the mark in the rear is made before the mark in the front. The very same evidence that convinces Alice, using the train frame, that the marks are made simultaneously, convinces Bob, using the track frame, that they are not made simultaneously. *Whether or not two events at different places happen at the same time has no absolute meaning—it depends on the frame of reference in which the events are described.* (Notice, in passing, that if you interchange the two words "place" and "time" in this sentence, the shocking assertion becomes quite humdrum: "Whether or not two events at different *times* happen at the same *place* has no absolute meaning—it depends on the frame of reference in which the events are described.")

People using the train frame, for whom the marks are made simultaneously, can use the arrival of the light signals to synchronize clocks at the front and rear of the train. Since people in the track frame maintain that the mark in the rear is made *before* the mark in the front, the track people would also maintain that the synchronization procedure used by the train people had actually resulted in the clock in the front of the train being behind the clock in the rear.

It is easy to find a precise quantitative measure of these disagreements. Let's analyze Alice's procedure for making marks simultaneously at both ends of the train, from the point of view of Bob in the track frame, where the train moves with speed v. It is convenient to call the length of the train L. I emphasize that by L we mean the length of the train *in the track frame*. Although we are used to assuming that the length of an object is independent of the frame in which it is measured, we can no longer take this for granted and, as noted earlier, we will indeed find it to be a false assumption.

In part 1 of figure 5.1, the light is turned on in the middle of the train, and the two pulses of light—which we shall again call photons—start moving from the center toward the front and the rear.

Part 2 of figure 5.1 shows things a time T_r later, just as the rearward moving photon meets the rear of the train, which has been moving toward it. At the instant of encounter a mark is made at the place on the tracks where the meeting takes place. During the time T_r the photon (which

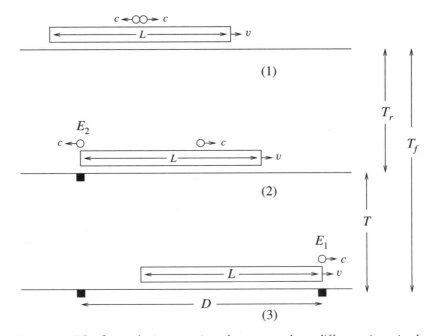

Figure 5.1. The figure depicts a series of events at three different times in the track frame. All lengths, times, and speeds are track-frame lengths, track-frame times, and track-frame speeds. The long horizontal rectangle is a train of length L moving to the right with speed v. The white circles are photons that move with speed c. (1) Two photons originate in the center of the train, moving toward the front and the rear. (2) At a time T_r after the first picture, the photon on the left reaches the rear of the train and a spot (black square) is left upon the tracks to mark the place where this happened (event E_2). (3) At a time T_f after the first picture, the photon on the right reaches the front of the train and another spot is left upon the tracks to mark the place (event E_1). The spot at E_1 is a distance D away from the spot at E_2. The time between the events E_1 and E_2 is $T = T_f - T_r$. As explained in the text, the time T and distance D between the two events E_1 and E_2 are related by $T = Dv/c^2$.

moves with speed c) has covered a distance cT_r. That distance is just half the length of the train, reduced by the distance the rear of the train (which is moving with speed v) has moved toward the photon in the time T_r. So

$$cT_r = \tfrac{1}{2}L - vT_r. \tag{5.1}$$

Part 3 of figure 5.1 shows things a (longer) time T_f after the light was turned on in part 1. At this moment the forward moving photon meets the front of the train, which has been moving away from it. At the instant of encounter a mark is made at the place along the tracks where the meeting takes place. During the time T_f the photon has covered a distance cT_f. But that distance is just half the length of the train, increased by the distance the front of the train has moved away from the photon in the time T_f. So

$$cT_f = \tfrac{1}{2}L + vT_f. \tag{5.2}$$

We want to find the time $T = T_f - T_r$ between the making of the two marks, so it is natural to subtract the second equation from the first, since the left side then becomes $c(T_f - T_r)$, which is just cT. A second advantage of this procedure is that the unknown length L disappears from the result, which is simply

$$cT = v(T_f + T_r). \tag{5.3}$$

But what is $T_f + T_r$? Fortunately this quantity times c has a very simple meaning: $c(T_f + T_r)$ is just the sum of cT_r, the distance light travels along the track from the place where the lamp was turned on (shown in part 1 of the figure) to the place on the track where it reaches the rear and the track is marked (shown in part 2). And cT_f is the distance light travels in the other direction along the track from the place where the light was turned on (shown in part 1) to the place on the track where it reaches the front and the track is marked (shown in part 3). Thus the total distance D along the track between the two marks is given by

$$D = c(T_f + T_r). \tag{5.4}$$

Replacing $(T_f + T_r)$ in (5.3) by D/c and dividing both sides of (5.3) by another factor of c so that T stands by itself on the left side, we have a relation between the track-frame time T between the making of the two marks and the track-frame distance D between them:

$$T = \frac{Dv}{c^2}. \tag{5.5}$$

We can abstract this into a general rule, by eliminating the talk of Alice, Bob, trains, tracks, and marks:

If events E_1 and E_2 are simultaneous in one frame of reference, then in a second frame that moves with speed v in the direction pointing from E_1 to E_2, the event E_2 happens at a time Dv/c^2 earlier than the event E_1, where D is the distance between the two events in the second frame.

How big an effect is this? Suppose the two marks are 10,000 feet of track apart, and suppose the speed of the train is 100 f/sec. Since the speed of light is a billion f/sec, Dv/c^2 works out to $10,000 \times 100/(1,000,000,000)^2 =$ one trillionth of a second (one *picosecond*). The two events that are simultaneous in the train frame are a trillionth of a second apart in the track frame. Not the sort of thing you'd be likely to notice. On the other hand, people who work with lasers these days are used to dealing with times a thousand times smaller than a picosecond (a *femtosecond*).

I remarked earlier that if you interchange time and space, the surprising fact that two frames of reference can disagree about whether two events in different places are simultaneous turns into the commonplace fact that two frames of reference can disagree about whether two events at different times happen in the same place. If we measure time in nanoseconds and distances in feet (or use any other units in which $c = 1$), then this intriguing symmetry under the interchange of time and space becomes quantitative as well as qualitative. The rule for simultaneous events says the following:

If two events happen at the same *time* in the train frame, then in the track frame the *time* between them in *nanoseconds* is equal to the *distance* between them in *feet*, multiplied by the speed v of the train along the tracks (in feet per nanosecond).

Take that statement and interchange time and space in every italicized word, making no other changes. What you get is a second rule:

If two events happen at the same *place* in the train frame, then in the track frame the *distance* between them in *feet* is equal to the *time* between them in *nanoseconds*, multiplied by the speed v of the train along the tracks (in feet per nanosecond).

This second rule is nothing more than the precise quantitative formulation of the familiar rule for how far something with a given speed—namely, the place on the train where the two events occur—moves in a given time. We shall come upon other examples of this wonderful symmetry between time and space.

The quantitative rule we have just found for simultaneous events gives rise to a similar quantitative rule for synchronized clocks. We have noted before that disagreements about whether two events are simultaneous are intimately related to disagreements about whether two clocks are synchronized. To formulate a quantitative rule about clock-synchronization disagreements, notice the following:

Suppose the times of the two markings are recorded in the track frame by two clocks, properly synchronized in the track frame and attached to the tracks at the places where the marks are made. How do people using the train frame, for whom the two marks are made simultaneously, account for the fact that the track-frame clocks read times that differ by

Dv/c^2? Easily! They say that the reason the track-frame clocks indicate the rear mark was made a time Dv/c^2 before the forward mark is that the track-frame clock that recorded the time the rear mark was made is actually *behind* the track-frame clock that recorded the time the forward mark was made by exactly that amount: Dv/c^2. This gives the following rule:

If two clocks are synchronized and separated by a distance D in their proper frame, then in a frame in which the clocks move along the line joining them with speed v, the reading of the clock in front is behind the reading of the clock in the rear by Dv/c^2.

Since $c = 1$ f/ns, a neat way to say this is that the clock in front is behind the clock in the rear by v nanoseconds per foot of proper-frame separation, where v is the speed of the clocks in f/ns. With v less than 1 f/ns, this isn't an enormous effect. It's less than a microsecond per 1,000 feet, and substantially less, if v is substantially less than the speed of light. If v is the speed of sound (a foot per millisecond) it's only a millionth of a microsecond per 1,000 feet, or a nanosecond per million feet.

Note that the $T = Dv/c^2$ rule for simultaneous events and the $T = Dv/c^2$ rule for synchronized clocks both relate a time T and a distance D *in one and the same frame*. In the rule for simultaneous events, the relation is between the distance D and the time T between the events in a frame in which they are not simultaneous. In the rule for synchronized clocks, D is the distance between the clocks in the frame in which they are synchronized, and T is the difference in their readings in the frame in which they are not synchronized, but those readings define "time" in the frame in which they *are* synchronized.

In chapter 4 we found that various ways of trying to produce an object moving faster than light—multistage rockets, or high-speed collisions—all failed to work. The $T = Dv/c^2$ rule for simultaneous events gives direct evidence of the very problematic nature of anything moving faster than light. Consider an object moving east along the tracks at a speed u greater than c. Events in the history of the object that happen when it is in two different places, a distance D apart in the track frame, will happen a time $T = D/u$ apart in the track frame. If we define the velocity v to be c^2/u (which is less than the speed of light, since c/u is less than 1), then we have $T = Dv/c^2$.

We can now use the rule for simultaneous events in the other logical direction. Since the time and distance between the two events are related by $T = Dv/c^2$ in the track frame, then in the frame of a train moving to the right at speed v the events must be simultaneous. In the frame of the train the object is in two different places at the same time!

This is such a bizarre situation that one's suspicion is strengthened that the difficulty we have already encountered in producing an object moving faster than light must be a reflection of the impossibility of such motion. This conclusion will be even more strongly supported by further problematic features of faster-than-light motion that we will be in a better position to examine in subsequent chapters.

In the remainder of this chapter I show that the $T = Dv/c^2$ rule for simultaneous events remains valid (as it must!) even when Alice's signals go from the middle to the ends of the train at a train-frame speed that is not the speed of light. The only reason for using light signals is that the analysis is substantially simpler, since light signals (and only light signals) have the same speed in both the train and track frames. While I would encourage you at least to glance at this discussion, it is not essential for the developments that follow.

The $T = Dv/c^2$ rule for simultaneous events is such a fundamental part of relativity, that it is worth emphasizing that it does not depend on Alice using light signals to ensure that the marks on the tracks are made simultaneously at the two ends of the train. She can use any two signals she likes, as long as they both travel from the middle of her train to the front and rear at the same train-frame speed u.

To see this, suppose the speed u of Alice's two signals, while not as fast as the speed of light, is greater than the speed v of her train in the track frame, so her signal to the rear of the train moves in the direction opposite to her signal to the front, even in the track frame. (A very similar argument can be constructed if u is less than v, in which case both signals move in the same direction in the track frame.) Let the track-frame speeds of Alice's signals to the front and rear be w_f and w_r. Later we shall construct these out of u and v using the velocity addition law of chapter 4. Figure 5.2 describes the situation. Note that it differs from figure 5.1 only in the replacement of c by w_f or w_r, depending on the direction of motion of the signal. As a result c must be replaced by the appropriate w in (5.1) and (5.2), which become

$$w_r T_r = \tfrac{1}{2}L - vT_r, \qquad (5.6)$$

and

$$w_f T_f = \tfrac{1}{2}L + vT_f, \qquad (5.7)$$

and the total distance D along the tracks between the two marks is now given by

$$D = w_f T_f + w_r T_r. \qquad (5.8)$$

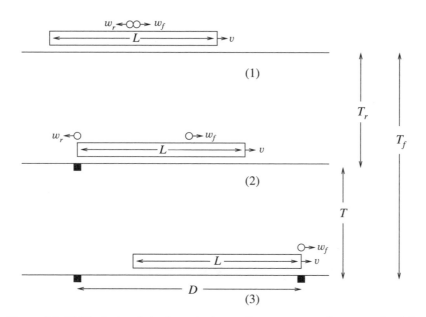

Figure 5.2. If Alice's signals in the train frame do not move to the two ends of the train with the speed of light c, then their speed in the track frame will also not be c. Figure 5.1 must therefore be relabeled. The white circles still represent Alice's signals, but they are no longer photons. One signal moves toward the front of the train with speed w_f. The other moves toward the rear with speed w_r.

To find the relation between the track-time time $T = T_f - T_r$ between the making of the two marks and the track-time distance D between the marks, we must now work harder than we had to when w_r and w_f were both c. First we solve (5.6) and (5.7) for T_r and T_f:

$$T_r = \frac{\frac{1}{2}L}{w_r + v}, \tag{5.9}$$

$$T_f = \frac{\frac{1}{2}L}{w_f - v}. \tag{5.10}$$

These give us T in terms of L:

$$T = \tfrac{1}{2}L\Big(\frac{1}{w_f - v} - \frac{1}{w_r + v}\Big). \tag{5.11}$$

When substituted into (5.8), they also give us D in terms of L:

$$D = \tfrac{1}{2}L\left(\frac{w_f}{w_f - v} + \frac{w_r}{w_r + v}\right). \tag{5.12}$$

Dividing (5.11) by (5.12) gives us a rather unwieldy expression for the ratio of T to D:

$$\frac{T}{D} = \frac{2v - (w_f - w_r)}{2w_f w_r + v(w_f - w_r)}. \tag{5.13}$$

Notice a few things about (5.13):

If Alice's signals *had* been light signals, then their speeds w_f and w_r in the track frame would both have been c. You can easily check that when $w_f = w_r = c$, (5.13) simplifies to $T/D = v/c^2$, giving us a much less graceful version of our earlier derivation of the $T = Dv/c^2$ rule.

Suppose, next, that the nonrelativistic velocity addition law were valid. We would then have

$$w_f = u + v, \quad w_r = u - v \tag{5.14}$$

so that $w_f - w_r = 2v$, and (5.13) becomes simply $T/D = 0$: the time between the making of the two marks is zero in the track frame as well as the train frame. This is indeed the nonrelativistic rule for simultaneous events: two events that are simultaneous in the train frame are also simultaneous in the track frame.

Finally, note that to get the correct relativistic result, we must replace the incorrect nonrelativistic relations (5.14) with the relativistic forms we found in chapter 4:

$$w_f = \frac{u + v}{1 + \frac{u}{c}\frac{v}{c}}, \quad w_r = \frac{u - v}{1 - \frac{u}{c}\frac{v}{c}}. \tag{5.15}$$

If you replace w_f and w_r in (5.13) with their correct relativistic expressions (5.15) in terms of u and v, and very carefully simplify the cumbersome expression that results, you will discover that the numerator of (5.13) differs from the denominator by nothing more than a factor of v/c^2, so that (5.13) reduces simply to

$$T/D = v/c^2. \tag{5.16}$$

We have thus recovered the $T = Dv/c^2$ rule, as long as Alice's signals both travel in the train frame with the same speed u, whether or not u is equal to the speed of light c.

So while it makes for a simpler argument, the constancy of the speed of light need not play a central role in a clock-synchronization procedure (though it did, of course, play a central role in the derivation of the relativistic velocity addition law that we used to find the speed of Alice's signals in the track frame.)

Six

Moving Clocks Run Slowly; Moving Sticks Shrink

IN CHAPTER 5 WE CONCLUDED that if two clocks are synchronized and separated by a distance D in a frame in which they are both at rest, then in a frame in which they move with speed v along the line joining them, they are not synchronized: the reading of the clock in front lags behind the reading of the clock in the rear by an amount T given by

$$T = Dv/c^2. \tag{6.1}$$

We deduced this rule by considering what Alice, who uses the train frame, has to say about the simultaneous reading of two clocks used by Bob, in the track frame. According to Alice, Bob's clocks are clearly out of synchronization since they assign different times to the simultaneous production of two marks on the track. Bob, of course, maintains that his two clocks are indeed synchronized, and that the correct conclusion is that the two marks on the track are not made simultaneously.

The principle of relativity assures us that the rule (6.1) must be valid in any inertial frame of reference, so although we deduced (6.1) by considering what Alice would say about two clocks synchronized by Bob in the track frame, the same rule must also apply to the discrepancy in readings that Bob attributes to two clocks synchronized by Alice in the train frame.

By exploring the consequences of (6.1) for two clocks synchronized by Alice in the train frame and two other clocks synchronized by Bob in the track frame, we can make some further deductions about the rate at which Alice maintains that Bob's clocks are running (and vice versa) or about the *length* that Bob assigns to Alice's train, or that Alice assigns to a given stretch of track. We shall deduce that moving clocks must run slowly and that moving trains (or moving tracks) must shrink along the direction of their motion, and we shall find the quantitative relation between the speeds of clocks or objects and the factors by which they slow down or shrink.

Let Alice's two synchronized clocks be attached to either end of her train. This is pictured from the point of view of the train frame in the right half of figure 6.1. Because both clocks are synchronized in the train frame, they both read the same time: 0. The figure also indicates that the *proper length* of the train—its length in its proper frame—is L_A. Also pictured are two clocks attached to the track, which move to the left with the track at speed v. The clocks on the track have been synchronized in the track frame. It follows that in the train frame the track clock in the front lags behind the track clock in the rear by an amount that we can call T_B.

On the other hand, because Alice's clocks are synchronized in the train frame, they are not synchronized in the track frame. This is pictured in the left half of figure 6.1. Because the train clocks are *not* synchronized in the track frame, it now requires *two* pictures taken at two different track-frame times to depict both of Alice's clocks reading 0. In the upper left picture, the clock at the rear of the train reads 0, and the clock in the front is behind the clock in the rear, reading a negative time that we can call $-T_A$. In the lower left picture, the clock in the front of the train has advanced from $-T_A$ to 0, while the clock in the rear has advanced by the same amount from 0 to $+T_A$. (They have advanced by the same amount because they are identical clocks moving at the same speed.) The track-frame time between the two pictures on the left is the time T_B that the two clocks attached to the tracks have advanced. That these two clocks are synchronized in the track frame is evident from either of the pictures on the left.

The quantitative rule (6.1) tells us that the amount T_A by which the two train-frame clocks differ in the track frame is related to the train-frame distance L_A between them by

$$T_A = L_A v/c^2, \tag{6.2}$$

where v is the speed of the train in the track frame. By the same token, the amount T_B by which the two track-frame clocks differ in the train frame is related to the track-frame distance D_B between them by

$$T_B = D_B v/c^2, \tag{6.3}$$

where (the same) v is the speed of the tracks in the train frame. From (6.2) and (6.3), together with a few elementary features of figure 6.1, we can deduce everything we need to know about the rates of moving clocks and the lengths of moving trains or tracks.

The slowing-down factor for moving clocks is given by T_A/T_B. (I call it a "slowing-down factor" because it turns out to be less than 1. If it turned out to be bigger than 1 it would have been a speeding-up factor.) This follows from the two track-frame pictures on the left of figure 6.1. Between the pictures both track-frame clocks advance by a time T_B, while both train-frame clocks advance by a time T_A. Since the track-frame clocks give correct time in the track frame, T_A is the time a train-frame clock advances in a track-frame time T_B. So the ratio T_A/T_B does indeed tell

Track Frame (Bob's)   **Train Frame (Alice's)**

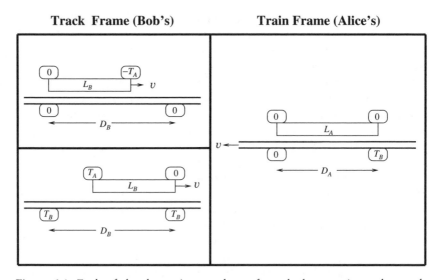

Figure 6.1. Each of the three pictures shows four clocks, a train, and a track. The train is the long rectangle. Two of the clocks are attached to it, one at the front, the other at the rear (the small rounded rectangles just above the front and rear of the train). The tracks are the two long parallel lines below the train. The other two clocks are attached to the tracks (the two small rounded rectangles shown below the tracks). The clocks attached to the train are synchronized in the train frame; those attached to the tracks are synchronized in the track frame. The time shown by a clock is indicated by the symbol inside it. The picture on the right depicts a single moment of time in the train frame. Both train clocks read the same time 0. The track and its attached clocks move to the left with speed v. The track clocks are not synchronized in the train frame: the clock in the front is behind the clock in the rear by a time T_B. The length of the train in the train frame is its proper length, L_A. The two clocks attached to the track are right next to the two clocks attached to the two ends of the train. It is evident from the figure that L_A is the same as D_A, the train-frame length of the segment of (moving) track that stretches between either pair of clocks. The two pictures on the left depict two different moments of time in the track frame. The first picture takes place when both track-frame clocks read 0; the second takes place when both read T_B. The train and its attached clocks move to the right with speed v. Note that the train-frame clocks are not synchronized in the track frame: the clock in front is behind the clock in the rear by a time T_A. The distance between the clock attached to the tracks in the track frame is the proper length D_B of the segment of track between them. The length of the train in the track frame is L_B.

us the fraction of track-frame time by which the train-frame clocks have advanced.

In the same way, the shrinking factor for a moving object is given by the ratio, L_B/L_A, of the track-frame length L_B of the train to its length L_A (its proper length) in the train frame, in which it is at rest. (If this ratio turned out to be a number bigger than 1 it would be a stretching factor, but I'm again anticipating the fact that it turns out to be less than 1.) The same shrinking factor is also given by the ratio D_A/D_B of the train-frame length D_A of the moving track between the two track-frame clocks, and the proper length D_B of that same stretch of track. So we have

$$L_B/L_A = D_A/D_B. \qquad (6.4)$$

To deduce the actual amount of slowing down and shrinking we need to note two other things.

1. It is evident from the train-frame picture on the right of figure 6.1 that the train-frame length D_A of (moving) track connecting the two track-frame clocks is equal to the train-frame (proper) length L_A of the train:

$$L_A = D_A. \qquad (6.5)$$

It is crucial for this conclusion that the train-frame synchronized clocks at the two ends of the train both read the same time, thereby revealing that the picture shows a *single moment* of train-frame time. Were the figure, on the contrary, a composite stitched together from fragments taken at different moments of train-frame time, then since the tracks are moving in the train frame, different parts of the tracks would be pictured at the places they occupied at *different moments* of train-frame time and we could conclude nothing about the train-frame length of the track between the clocks.

2. The track-frame pictures on the left give a slightly more complicated relation between L_B and D_B. According to those pictures, D_B is the track-frame distance between the left end of the train at track-frame time 0 and the right end at track-frame time T_B. This distance is given by the track-frame length L_B of the train augmented by the distance the train moves between the two pictures. Since the track-frame time between the two pictures is T_B (as indicated by the track-frame clocks in the figure) and since the train moves with speed v, that additional distance is vT_B. So we have

$$D_B = L_B + vT_B. \qquad (6.6)$$

Everything we want follows from the relations (6.2)–(6.6). To begin with, we can conclude immediately from (6.2), (6.3), and (6.5) that the slowing-down factor for moving clocks must be *the same* as the shrinking factor for moving objects. For (6.2) and (6.3) tell us that $T_A/T_B = L_A/D_B$, while (6.5) tells us that $L_A = D_A$. Consequently

$$T_A/T_B = D_A/D_B. \tag{6.7}$$

The left side of this relation is the slowing-down factor for moving clocks; the right side is the shrinking factor for moving objects. Calling this common factor s (which, conveniently, works for either "shrinking" or "slowing-down"—for that matter it would also have worked for "stretching" or "speeding-up"), we can write

$$T_A = sT_B, \quad D_A = sD_B, \quad \text{and also} \quad L_B = sL_A \tag{6.8}$$

(where the last of these follows from (6.4)).

To learn how the the shrinking (slowing-down) factor s, depends on v, note that if we combine (6.6) with (6.3), we find that $D_B = L_B + v^2 D_B/c^2$, which tells us that

$$L_B = D_B(1 - v^2/c^2). \tag{6.9}$$

But (6.8) tells us that $L_B = sL_A$, (6.5) tells us that $L_A = D_A$, and (6.8) tells us that $D_A = sD_B$. Putting these three together tells us that $L_B = s^2 D_B$, and therefore (6.9) tells us that

$$s^2 D_B = D_B(1 - v^2/c^2). \tag{6.10}$$

Consequently the shrinking factor (or slowing-down factor) is

$$s = \sqrt{1 - v^2/c^2}. \tag{6.11}$$

(In the literature on relativity the quantity $1/s$ is often denoted by γ. For our purposes $s = 1/\gamma$ is more intuitive and will be more useful.)

Note that s is the square root of a number less than 1, so s is indeed less than 1 and is indeed a shrinking (not stretching) or slowing-down (not speeding-up) factor. Note also that if v were to exceed the speed of light c, then (6.10) would make s^2 a negative number, which makes no sense. Indeed, the analysis in chapter 5 only makes sense if the speed v of the train along the tracks is less than speed of light c, for if the train moved faster than light, then the photon from the middle of the train would never reach the front of the train in the track frame. These are further hints that the speed of light might be an upper limit to how fast anything can be moving in any inertial frame of reference.

The shrinking of moving objects along the direction of their motion is called the Fitzgerald contraction, in honor of the otherwise little-known Irish physicist who first suggested it. It is also called the Lorentz-Fitzgerald contraction, in honor of the great Dutch physicist H. A. Lorentz, who had the same idea at about the same time. Often it is just called the Lorentz contraction, a manifestation of the unfortunate Matthew effect. ("To him that hath shall be given; from him that hath not shall be taken even that which he hath.") Lorentz and Fitzgerald suggested that objects should experience such a contraction when moving relative to "the ether." They missed the insight of Einstein that the contraction must take place in *any* frame of reference for objects that move in that frame of reference, and they missed Einstein's insight into the frame-dependent nature of simultaneity and the essential role it plays in tying everything together into a consistent picture.

The slowing down of moving clocks is often referred to by the deplorable term "time dilation." It is deplorable because it suggests in some vague way that "time itself" (whatever that might be) is expanding when a moving clock runs slowly. While the notion that time stretches out for a moving clock has a certain intuitive appeal, it is important to recognize that what we are actually talking about has nothing to do with any overarching concept of time. It is simply a relation between two sets of clocks. While it is commonly believed that there is something called time that is measured by clocks, one of the great lessons of relativity is that the concept of time is nothing more than a convenient, though potentially treacherous, device for summarizing compactly all the relationships holding between different clocks. If one set of clocks is considered to be stationary, synchronized, and running at the correct rate, then a second set, considered to be moving (and synchronized in the frame in which they are all stationary) will be found to be both asynchronized and running slowly, according to the first set. But if we consider the second set to be stationary, synchronized, and running at the correct rate, then according to the second set the first set will be found to be asynchronized and running slowly. Chapter 9 illustrates this in detail.

In both cases—the shrinking of moving sticks or the slowing down of moving clocks—one is inclined to be skeptical of these conclusions. How can Alice maintain that Bob's clocks are running slowly, and Bob maintain that Alice's clocks are running slowly, when they are both talking about the same sets of clocks? If Alice maintains that Bob's clocks are running slowly, shouldn't Bob necessarily maintain that Alice's clocks are running *fast*? Similarly for sticks, lined up along their direction of relative motion. If Alice maintains that Bob's moving sticks have shrunk compared with her stationary sticks, then shouldn't Bob have to maintain that Alice's moving sticks have *stretched* compared with his stationary sticks?

To succumb to these doubts is to forget that Alice and Bob also disagree on whether two events in different places happen at the same time or, equivalently, on whether two clocks in different places are synchronized. Because of this fundamental disagreement, each of them maintains that the other has determined the rate of moving clocks or the length of moving sticks *incorrectly*. For to measure the length of a *moving* stick, you must determine where its two ends are *at the same time*. If you don't get the locations of the two ends at the same time, then the stick will have moved between your two determinations of where its ends are, and you won't be getting its length right. But to do this requires a judgment about whether spatially separated events at the two ends of the stick are simultaneous. Similarly, to compare how fast a moving clock is running with the rate at which stationary clocks run, it is necessary to compare at least two of the readings of the moving clock with the readings of the stationary clocks that are next to it when it shows those readings. But since the moving clock *moves*, this requires one to use two *synchronized* stationary clocks that are in two *different* places.

There is thus nothing inconsistent in Alice and Bob each saying that the other's clocks are running slowly and each saying that the other's sticks have shrunk. Each can point to a flaw—a failure to use properly synchronized clocks—in the procedure that the other uses to make such determinations. This is not, however, to say that the phenomena of "time dilation" and length contraction are mere conventions about how we use language to describe the behavior of clocks and measuring sticks. As we shall see, these phenomena can have quite striking consequences. These consequences can be observed in any frame of reference, although the *explanations* given for the consequences can differ substantially from one frame of reference to another.

One of the simplest manifestations of such behavior, which has been known and observed for over half a century, is provided by unstable sub-atomic particles. These have a characteristic lifetime τ. If you have a group of such particles, at rest or moving slowly, about half of them disintegrate within a time τ. Their collective statistical behavior therefore provides a kind of clock. Furthermore, in contrast to watches or alarm clocks, sub-atomic particles can be accelerated up to speeds u very close to the speed of light. When a group of such particles is moving along a track at a speed u close to the speed of light c, most of the particles in the group manage to travel without disintegrating over distances of track very much greater than the typical distance $u\tau$ one would expect them to be able to cover if their survival rate were unaffected by their motion. They can go much further because their "internal clocks" that govern when they decay are running much more slowly in the frame in which they rush along at speeds close to c. This is a real effect, and it plays a crucial role in the operation of such particle accelerators.

But how can this behavior be reconciled with the fact that in the frame moving with the particles, their internal clocks are running at the normal rate, and only about half of them can survive for a time τ? The reconciliation comes from the fact that in the frame that moves with the particles, the track along which they move is rushing by at the speed u. All distances along the track are therefore reduced by the shrinking factor, and much more of the track can go past the particles in the time τ than could have gone past if the track had not shrunk.

So both frames agree that half the particles are able to cover a greater length $u\tau/s$ of track, where s, the shrinking or slowing-down factor for things moving with the velocity u of the particles, can be very small if the velocity is close to c. In the track frame this is because a typical particle survives for a time τ/s, which is much longer than the time τ it would survive if it were stationary. In the rest frame of the particles it is because the length of the track has shrunk by a factor s, so the length of moving track that can get past the particle in the time τ is augmented by the factor $1/s$. The explanatory stories differ, but the resulting behavior is the same.

This effect was observed in the behavior of particles called μ-mesons, even before the age of powerful particle accelerators. The μ-mesons are produced by cosmic rays in the upper atmosphere. When at rest they have a lifetime of about 2 microseconds, so if their internal clocks ran at a rate independent of their speed, even if they traveled at the speed of light about half of them would be gone after they had traveled 2,000 feet. Yet about half the μ-mesons produced in the upper atmosphere (about 100,000 feet up) manage to make it all the way down to the ground. This is because they travel at speeds so close to the speed of light that the slowing down factor is $s = 1/50$, and they can survive for 50 times as long as they can when stationary. In the frame of the μ-mesons, of course, their lifetime remains 2 microseconds, but half of them still make it down to the ground because the earth is rushing up at them so fast that the height of the atmosphere contracts by a factor of $1/s = 50$, from 100,000 feet to 2,000 feet.

Note that although there are substantial disagreements between Alice's train-frame picture of events and Bob's track-frame picture, whenever the two pictures are narrowed down to describe only things that happen *both* in the same place *and* at the same time, the restricted pictures agree. This is illustrated in figure 6.2, in which figure 6.1 is redrawn to emphasize the space-time coincidences that Alice and Bob agree on. This is quite a general state of affairs. All frames of reference will agree in their description of space-time coincidences—events that happen both at the same time and

Track Frame (Bob's)   **Train Frame (Alice's)**

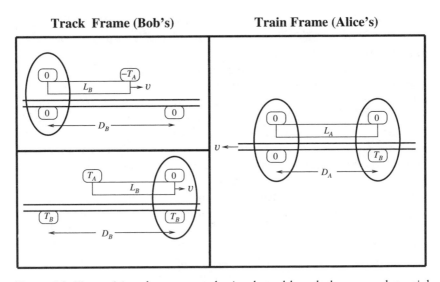

Figure 6.2. Figure 6.1, redrawn to emphasize that, although there are substantial differences of opinion between Alice and Bob, there is complete agreement about things that happen at the same place *and* at the same time. Thus the events that are encircled on the left of the upper track-frame picture (two clocks being together at the rear of the train and reading 0) are described in exactly the same way in the encircled region on the left of the train-frame picture. And the two events that are encircled on the right of the lower track-frame picture (two clocks being together at the front of the train, the clock on the train reading 0 and the clock on the track reading T_B) are also described in exactly the same way in the encircled region on the right of the train-frame picture. There is a disagreement about whether the two events *are* (train frame) or *are not* (track frame) simultaneous. And there is a lot of disagreement about what is going on *somewhere else* while the encircled events are happening. Note that "while" means "at the same time as" and is therefore a potentially treacherous word.

in the same place. Differences of opinion only arise when it comes to "stitching" together such events to tell a more elaborate story of what things are like everywhere at a given time. The disagreements arise because "at a given time" means different things in different frames. When this is fully taken into account, the disagreements are revealed merely as different conventional ways of describing the same phenomena.

There is a simple reason why the slowing-down factor s for moving clocks must be the same as the shrinking factor s for moving sticks, which makes no use of the Dv/c^2 rule for simultaneous events. The reason is that if the slowing-down factor and shrinking factors were different, then the behavior of the moving μ-mesons would come out differently, depending on whether you calculated it in the rest frame of the mesons or the rest

Stick Frame   **Clock Frame**

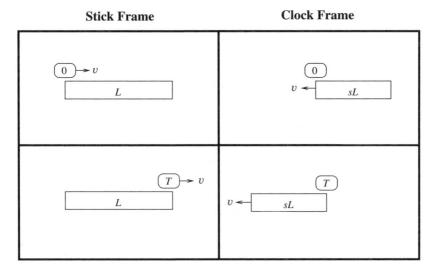

Figure 6.3. A stick and a clock, in relative motion. On the left we see things as described in the rest frame of the stick. The (proper) length of the stationary stick is L. The clock moves with speed v from the left end of the stick to the right end. It reads 0 when it passes the left end of the stick and T when it passes the right end. On the right we see things as described in the rest frame of the clock. The stick moves to the left with speed v and its length is sL, where s is the shrinking factor for moving sticks. The clock is stationary. It reads 0 when the left end of the stick passes it, and T when the right end of the stick passes it.

frame of the Earth. Here is a slightly more abstract way to make the same point:

Suppose (see figure 6.3) we have a stick of proper length L along which a clock moves to the right with speed v. Let the clock read 0 when it is at the left end of the stick and T as it passes the right end. This is depicted in the stick-frame in the two pictures on the left of figure 6.3, and in the clock-frame in the two pictures on the right.

If s is the shrinking factor for moving sticks, then in the clock frame we have a stick of length sL moving with speed v to the left. The time it takes the full length of the stick to get past the clock is just the time it takes the stick, moving at speed v, to go its full length sL: $T = sL/v$. Since the clock tells correct time and reads 0 when the left end of the stick passes it, it must read $T = sL/v$ when the right end passes it.

In the stick frame, on the other hand, the time it takes the clock to go from the left to the right end of the stick is L/v, the unshrunken length L of the stick divided by the speed v of the clock. Since the clock runs slowly in the stick frame, during this time the reading of the clock advances only by $T = s'L/v$, where s' is the slowing-down factor for moving clocks, and this is what it reads as it passes the right end of the stick.

But both frames must agree on the time T shown on the clock when it is at the right end of the stick, since this judgment is about things that happen at a single place and time. Therefore the slowing-down factor s' for moving clocks must be the same as the shrinking factor s for moving sticks.

We have now found all the basic relativistic facts about clocks and measuring sticks.

RULE FOR SYNCHRONIZED CLOCKS

If two clocks are stationary, synchronized, and separated by a distance D in one frame of reference, then in a second frame, in which they are moving with speed v along the line joining them, the clock in front lags behind the clock in the rear by

$$T = Dv/c^2. \tag{6.12}$$

RULE FOR SHRINKING OF MOVING STICKS OR SLOWING DOWN OF MOVING CLOCKS

The shrinking (or slowing-down) factor s associated with a speed v is given by

$$s = \sqrt{1 - v^2/c^2}. \tag{6.13}$$

Note a possible source of confusion: a clock moving with speed v runs slowly by a factor s, but whether you should multiply or divide by the slowing-down factor depends on what question you want to answer. The things to keep in mind are (a) that moving clocks run *slowly* and (b) that the slowing-down factor s is *less* than 1. So if the question is how much the reading on a moving clock advances in a time T, the answer is that it advances by sT, since the moving clock runs slowly, and multiplying T by s gives a number less than T. But if the question is how much time it takes for the reading on a moving clock to advance by T, the answer is T/s, since it takes a clock running slowly more time than T, and dividing T by a number s less than 1 produces a bigger number. Memorizing a formula doesn't help here. Thinking does. Similarly, a stick moving along its own direction with speed v shrinks by a factor s. So if the stick has proper length L, then when moving with speed v its length is only sL. On the other hand, if a moving stick has length L, then its proper length is L/s.

In the rest of this chapter I examine another way to synchronize clocks in different places, like those considered early in chapter 5 but rejected because we did not know the effect of its motion on the rate of a clock. What emerges from the irritatingly complicated algebra is nothing but the old $T = Dv/c^2$ rule that we found much more simply in chapter 5. While I would encourage you at least to glance at this discussion, which again illustrates the mutual consistency of the whole picture, it is not necessary for subsequent developments for you to follow the argument in detail.

Suppose Alice has two identical stationary clocks in the *same* place, so she can synchronize them unproblematically by direct comparison. After setting both clocks to 0, she sets one of them into uniform motion with speed u. (She can check that its speed is u without the problematic use of additional clocks by comparing its speed with that of a photon, as described in chapter 4.) The second clock is allowed to move away from the first until the moving clock reads t, when it stops moving. While the second clock moved it was running slowly, so it actually took (in Alice's frame) the longer time t/s_u, to make its journey, where s_u is the slowing-down factor $\sqrt{1 - (u/c)^2}$. This longer time is the amount by which the first clock advanced during the journey of the second, since it remained stationary and therefore continued to tell correct Alice-time. So to get the second clock back into synchronization with the first in Alice's frame, the moment it stops moving it must be reset from t to t/s_u.

This procedure results in two synchronized clocks in Alice's frame. The distance D between the clocks is the speed u at which the second clock moved, times the time t/s_u that it was moving:

$$D = ut/s_u. \tag{6.14}$$

Now consider how all this is described by Bob, who moves opposite to the direction of motion of the second clock with speed v. (The same general conclusion can be reached if Bob moves in the same direction as the second clock.) According to Bob, the first clock always moves with speed v, but the second clock, while the two are separating, moves with a higher speed w given by the relativistic velocity addition law

$$w = \frac{u + v}{1 + \left(\frac{u}{c}\right)\left(\frac{v}{c}\right)}. \tag{6.15}$$

Bob must agree with Alice that the second clock started to move away from the first when both read 0, and that its motion relative to the first continued until the second clock read t, since each of those judgments is about things that happen at the *same* place *and* time. Since the speed of the second clock was w during this period, Bob can use the second clock to determine how long the clocks were in relative motion, by dividing t by the slowing-down factor s_w. So according to Bob the two clocks were in relative motion for a time t/s_w.

Because the first clock is always moving with speed v in Bob's frame, it slows down by s_v. Therefore during the time t/s_w that the clocks are moving apart, the first clock advances from 0 to $s_v(t/s_w)$. This is what it reads at the moment the second clock ceases to move relative to the first, and is reset to t/s_u. After that moment both clocks are again moving at the same speed v and therefore running at the same rate, so from that moment on, the reading of the first clock will continue to differ from that of the second by

$$T = \left(\frac{s_v}{s_w} - \frac{1}{s_u}\right)t = \left(\frac{s_v s_u}{s_w} - 1\right)\frac{D}{u}, \qquad (6.16)$$

where (6.14) has been used to express t in terms of D.

To recover the Dv/c^2 rule from (6.16) requires nothing more than a rather cumbersome bit of algebraic manipulation. When w is given by the addition law (6.15), the slowing down factor s_w becomes

$$s_w = \sqrt{1 - (w/c)^2} = \sqrt{1 - \left(\frac{\frac{u}{c} + \frac{v}{c}}{1 + \left(\frac{u}{c}\right)\left(\frac{v}{c}\right)}\right)^2} = \frac{\sqrt{\left(1 + \left(\frac{u}{c}\right)\left(\frac{v}{c}\right)\right)^2 - \left(\frac{u}{c} + \frac{v}{c}\right)^2}}{1 + \left(\frac{u}{c}\right)\left(\frac{v}{c}\right)}$$

$$= \frac{\sqrt{\left(1 - \left(\frac{u}{c}\right)^2\right)\left(1 - \left(\frac{v}{c}\right)^2\right)}}{1 + \left(\frac{u}{c}\right)\left(\frac{v}{c}\right)} = \frac{s_u s_v}{1 + \left(\frac{u}{c}\right)\left(\frac{v}{c}\right)}. \qquad (6.17)$$

Consequently

$$s_u s_v / s_w = 1 + uv/c^2, \qquad (6.18)$$

so (6.16) does indeed assert that the reading of the first clock differs from that of the second by

$$T = Dv/c^2. \qquad (6.19)$$

So regardless of the speed u with which Alice separates the clocks in her frame, Bob will conclude that the clock in front (the second) is behind the clock in the rear (the first) by Dv/c^2, just as the $T = Dv/c^2$ rule requires. The rule does not depend on Alice using light signals (or any other signals) to synchronize her clocks.

There is an elegant minor refinement of Alice's procedure. The procedure requires the moving clock to be reset after it reaches the distance D from the stationary one, to correct for the fact that it ran slowly throughout its journey. But since the slowing-down factor $s_u = \sqrt{1 - u^2/c^2}$ is very close to 1 when u is small compared with the speed of light, you might think that by moving the clocks apart slowly enough Alice could get them into almost perfect synchronization without having to take this extra step. If, however, the two clocks have to end up a fixed distance D apart, then the time D/u that this takes will get very large as u gets very small. So although the slowing down is less important when u is small, it has more time to build up its effect, and it's not obvious that the total effect will be unimportant. But closer examination reveals that nearness of the slowing-down factor to 1 more than compensates for the longer duration of the journey. We can see this as follows:

At the end of its journey Alice's clock is reset from $t = (D/u)s_u$ to D/u. The difference between these two readings is

$$T_0 = D \left(\frac{1 - \sqrt{1 - u^2/c^2}}{u} \right). \tag{6.20}$$

We can make the dependence of T_0 on u, when u is very small, much clearer by multiplying numerator and denominator of (6.20) by $1 + \sqrt{1 - u^2/c^2}$. If we take advantage of the algebraic fact that $(a - b)(a + b) = a^2 - b^2$, we then have

$$T_0 = D \frac{u/c^2}{1 + \sqrt{1 - u^2/c^2}}. \tag{6.21}$$

When u is small compared with the speed of light, the denominator in (6.21) hardly differs from $1 + 1 = 2$, and the discrepancy between what the second clock reads at the end of its journey and what it should read is very close to

$$T_0 = \tfrac{1}{2}Du/c^2. \tag{6.22}$$

So by reducing the speed u with which the clocks separate, Alice can reduce the discrepancy T_0 by the same amount. By letting the clocks separate slowly enough, she can make the discrepancy as small as she pleases. Once the discrepancy becomes smaller than the accuracy with which she can read her clocks, she no longer needs to reset the second clock after it has stopped moving.

This procedure is called clock synchronization by slow transport. While it works perfectly well in Alice's frame, as we have already seen, when viewed in Bob's frame it results, regardless of the speed with which the clocks are separated, in their being out of synchronization almost exactly as required by the Dv/c^2 rule. I say "almost" because Bob would have found an exact discrepancy of Dv/c^2 had Alice changed the reading of the second clock by T_0. Because she no longer does this, the discrepancy Bob finds differs from Dv/c^2 by T_0. But since T_0 was chosen by Alice to be less than the accuracy with which any clock can be read, this is of no consequence for Bob.

Seven

Looking At a Moving Clock

WE HAVE ESTABLISHED THAT IN any inertial frame of reference a clock that moves with speed v runs slowly compared with stationary clocks. The slowing-down factor is given by

$$s = \sqrt{1 - v^2/c^2}. \qquad (7.1)$$

One may have the feeling that this is just some kind of trick—a conclusion based on playing intellectual games with the concept of simultaneity. If you actually *looked* at a moving clock would you actually *see* it running slowly?

The answer is that what you see depends on whether the clock is moving toward you or away from you. If it moves away from you, you do indeed see it running slowly, but considerably more slowly than the (slow) rate at which it actually is running. But if it moves toward you, you actually see it running fast!

This disparity between how fast a clock runs and how fast you *see* it run is a simple consequence of the fact that you do not *see* a clock reading a particular number until light that leaves the clock when it displays that number has traveled from the clock to your eyes. In thinking about this it is helpful to think of the clock as a digital clock that signals its reading by flashing a number. But of course even when you look at a conventional clock, the only reason you can see it is that light has bounced off its hands and then traveled to your eyes at the speed of light.

If the clock is standing still in your frame, the delay between the clock displaying a number and you seeing it display that number doesn't matter at all, because the extra time between the clock flashing each new number and the light actually reaching you is the same for each number. So although there is a delay before you see each flash, you receive the flashes at the same rate the clock is emitting them, and therefore you see the clock running at its actual rate, though behind what it actually reads in your frame at the moment you see it, because of the delay.

If the clock is moving away from you, the light from each successive flash has further to go before it reaches you, so you see the clock running more slowly than it actually is running in your frame. On the other hand, if the clock is moving toward you, the light from each successive flash has less distance to cover, so you see the clock running faster than it actually is running in your frame. It turns out that when the clock moves toward you, this effect is more important than the fact that the clock is running slowly, so you *see* it running fast.

It is not hard to construct a quantitative measure of this effect, which is called the *relativistic Doppler effect*. In fact it is possible to do so without knowing the actual value (7.1) of the slowing-down factor s. Indeed, the argument that follows provides, on the side, a derivation of the fact that $s = \sqrt{1 - v^2/c^2}$ that is independent of the analysis in chapter 6. Furthermore, it provides a derivation of the relativistic velocity addition law independent of the one given in chapter 4. And it doesn't even make use of the $T = Dv/c^2$ rules for simultaneous events or synchronized clocks used in chapter 5. So it could have been presented immediately after chapter 3, as an alternative route to developing the entire subject. Of course from a logical point of view there is no need for independent arguments leading back to conclusions we have already established. But in the case of relativity, where some of the conclusions are so strange, it can be reassuring to see that the same conclusions emerge from quite different lines of thought.

So take a clock that flashes a new number every T seconds in its proper frame. Let $f_t T$ and $f_a T$ be the number of seconds in your frame between the flashes that reach you when the clock moves toward (t) or away (a) from you with speed v. We shall deduce the values of f_a and f_t, the slowing-down and speeding-up factors for what you *see*, as well as the value of s, the slowing-down factor for what the clock is actually doing in your frame. In what follows, I anticipate the fact (which we already know, but are about to rederive) that s is less than 1—a moving clock runs slowly. But the same argument, with a few words suitably changed, would work just as well if we started off unsure of whether or not s were less than 1.

Since the moving clock runs slowly it only flashes a new number every T/s seconds. During that time it gets a distance $v(T/s)$ further from (or closer to) you, so the light from each successive flash takes a time $v(T/s)/c$ more (or less) to get to you. Consequently the time between light from the flashes reaching you (and therefore the time between your *seeing* successive flashes) is

$$f_a T = T/s + v(T/s)/c = (T/s)(1 + v/c) \tag{7.2}$$

if the clock moves away from you. And it is

$$f_t T = T/s - v(T/s)/c = (T/s)(1 - v/c) \tag{7.3}$$

if the clock moves toward you. Therefore

$$f_a = (1/s)(1 + v/c) \qquad (7.4)$$

and

$$f_t = (1/s)(1 - v/c). \qquad (7.5)$$

Since we already know the value of the slowing-down factor s from chapter 6, we are finished: we now know f_a and f_t. But even if we didn't know the value of s, we are now in a position to figure it out from the following neat idea:

Suppose Alice and Bob are stationary in the *same* frame of reference at different places, and Bob holds a clock that Alice watches. Suppose Bob's clock flashes every t seconds in its proper frame. Since Bob's clock is stationary with respect to Alice, every flash takes the same time to reach her. Since Alice's clock runs at the same rate as Bob's, Alice therefore sees a flash every t seconds according to her own clock. Now suppose that Carol moves from Bob to Alice at speed v. Each time Carol sees a new number appear on Bob's clock, she reinforces it with a flash of her own. She can do this automatically by adjusting the fast-slow setting on a clock moving with her so that it flashes a new number at the same rate that she sees Bob's numbers. Since Carol moves *away* from Bob with speed v, she sees a flash from Bob's clock every $f_a t$ seconds. She therefore adjusts her flasher to emit a number every T seconds, where $T = f_a t$. Since Carol and her flasher move *toward* Alice at speed v, Alice *sees* Carol's flasher flashing every $f_t T = f_t f_a t$ seconds. But since Carol's flashes arrive together with Bob's, and Alice sees one of Bob's flashes every t seconds, Alice must also see one of Carol's flashes every t seconds. So the effects of Carol seeing Bob's clock flash slowly and Alice seeing Carol's clock flash fast must cancel precisely:

$$f_t f_a = 1. \qquad (7.6)$$

When we combine (7.6) with (7.4) and (7.5), we learn everything of interest. Note first that (7.4) and (7.5) together require that

$$f_t f_a = (1/s)^2 (1 + v/c)(1 - v/c) = (1/s)^2 (1 - v^2/c^2). \qquad (7.7)$$

In view of (7.6) this immediately gives us an independent way of arriving at the form (7.1) of the slowing-down factor s. On the other hand (7.4) and (7.5) also tell us that

$$f_t/f_a = \frac{1 - v/c}{1 + v/c}. \tag{7.8}$$

Combining this with (7.6), which tells us that $1/f_a = f_t$, we immediately learn that

$$f_t = \sqrt{\frac{1 - v/c}{1 + v/c}}, \tag{7.9}$$

and therefore f_a (which is $1/f_t$) is given by

$$f_a = \sqrt{\frac{1 + v/c}{1 - v/c}}. \tag{7.10}$$

So we have derived f_t, f_a, and s without using any of our earlier results from chapters 4, 5, and 6!

To be concrete about it, suppose that $v = \frac{3}{5}c$, so the slowing-down factor is

$$\sqrt{1 - (\tfrac{3}{5})^2} = \tfrac{4}{5}. \tag{7.11}$$

This tells us that a clock moving at 60 percent of the speed of light takes $\frac{5}{4} = 1.25$ seconds to flash each second—it runs at $\frac{4}{5} = 80$ percent of its normal rate. But

$$\sqrt{\frac{1 - \frac{3}{5}}{1 + \frac{3}{5}}} = \tfrac{1}{2}, \tag{7.12}$$

so if the clock is moving toward you, you see it flash a new second every half second—i.e. you see it running at *twice* its normal rate. If it moves away from you, you see it flash a new second every 2 seconds—i.e. you see it running at *half* its normal rate. If $v = \frac{3}{5}c$, the slowing-down factor drops to $\frac{3}{5}$. But according to (7.9) and (7.10) the rate at which you *see* the clock flash differs from the rate at which it runs in its proper frame by a factor of 3.

A modest generalization of the argument leading to (7.6) gives us the relativistic velocity addition law in an entirely different way from how we derived it in chapter 4. Suppose Bob and Charles both move to the right, away from Alice, at speeds v and w in Alice's frame of reference, and

Charles moves to the right, away from Bob, at a speed u in Bob's frame. If Alice has a clock that flashes once a second, then Charles will receive a flash from her every $\sqrt{\frac{1+w/c}{1-w/c}}$ seconds, while Bob will receive a flash from Alice every $\sqrt{\frac{1+v/c}{1-v/c}}$ seconds. So if Bob reinforces the flashes he receives from Alice by setting his own clock to flash at the same rate he receives Alice's flashes, then Charles will receive the reinforcing flashes from Bob every $\sqrt{\frac{1+u/c}{1-u/c}}\sqrt{\frac{1+v/c}{1-v/c}}$ seconds. Since this must be the same as the rate at which Charles directly receives the flashes from Alice, we must have

$$\sqrt{\frac{1 + w/c}{1 - w/c}} = \sqrt{\frac{1 + u/c}{1 - u/c}} \sqrt{\frac{1 + v/c}{1 - v/c}}. \tag{7.13}$$

But (7.13) is entirely equivalent to the velocity addition law written in its multiplicative form,

$$\frac{c - w}{c + w} = \left(\frac{c - u}{c + u}\right)\left(\frac{c - v}{c + v}\right). \tag{7.14}$$

A similar effect, less noted (and less useful) than the Doppler effect, takes place when one looks at a moving train. Suppose you are standing right next to the tracks, watching an approaching train of proper length L. If the train is moving with speed v, its length in the track frame will be reduced to $s_v L$. But it will not look that short. Light from the rear of the train (imagine a lamp at the end of the train that you can see from the front) has to cover a greater distance before it reaches you than light from the front, and therefore at any moment the image of the rear of the train reaching your eyes will be from an earlier time, when the train was further away from you, than the image of the front reaching your eyes at that same moment. If you do some simple analysis, quite similar to what we did in deriving the relativistic Doppler effect, you will find that this more than compensates for the shrinking factor, and that the length the approaching train *is seen to have* is actually *longer* than its proper length by a factor

$$\sqrt{\frac{1 + v/c}{1 - v/c}}. \tag{7.15}$$

On the other hand if you watch a departing train, light from the front of the train has to cover a greater distance before reaching you than light from the rear. So at any moment the image of the front of the train reaching your eyes will be from an earlier time, when the train was closer to you, than the image of the rear reaching your eyes at that same moment. The train you *see* will therefore appear even shorter than its shrunken length. The same kind of analysis leads to the conclusion that the length the departing train appears to have is shorter than its proper length by the factor

$$\sqrt{\frac{1 - v/c}{1 + v/c}}. \tag{7.16}$$

I conclude this chapter with a puzzle: the preceding argument leads to the relativistic slowing-down factor (7.1) from nothing more than some simple considerations on how fast a clock looks to be running as it moves toward or away from you. The phenomenon of relativistic slowing down, however, depends crucially on the principle of the constancy of the velocity of light, which never appeared in the course of the argument. What's going on here?

The answer to the puzzle is that although that principle was never explicitly invoked, it sneaked into the argument in a rather subtle way. We derived the relations (7.4) and (7.5) in the frame of reference in which the person watching the clock was stationary. When we applied those relations to get (7.6), we used the factor f_a appropriate to Carol's frame and the factor f_t appropriate to Alice's. The appearance of one and the same value for c in both Carol's f_a and Alice's f_t only makes sense if the speed c of the flashes of light is the same in both Carol's frame and Alice's.

Eight

The Interval between Events

WE HAVE IDENTIFIED A VARIETY of things that people using different inertial frames of reference disagree about: the rate of a clock, the length of a
stick, whether two events are simultaneous, whether two clocks are synchronized. There are also some things on which people using different
frames of reference do agree: people in all frames of reference agree about
space-time coincidences—whether two events occur at both the same time
and the same place. And people in all frames of reference agree, of course,
about whether something moves at the speed of light c.

There are other things that people using different frames of reference
agree on. The constancy of the speed of light is, in fact, only a special case
of this broader group of quantities. Invariants—quantities that everybody
agrees on regardless of their frame of reference—play a more important
role in our understanding of the world than quantities that vary from
one frame of reference to another. The theory of relativity identifies such
invariant quantities, and in this sense "theory of relativity" is a terrible
name. "Theory of invariance" would have been better, since the most
important content of the theory is its identification of quantities that do
not change from frame to frame.

We can get a hint at what some of these new invariants might be, by first
giving a more abstract statement of the constancy of the velocity of light.
Consider two distinct events E_1 and E_2. Each event happens at a definite
place and time, although different frames of reference may disagree about
the place and the time at which each of the events occurs. Let D and T
be the distance and time between the two events according to a particular
frame of reference.

If the two events happen to be events in the history of a single photon
moving uniformly at speed c, then $D/T = c$. Since the speed of the photon
is the same in all frames of reference, in any other frame of reference
the distance D' and the time T' between E_1 and E_2 are also related by
$D'/T' = c$, even though D' need not be equal to D nor need T' be equal
to T. We can turn this into an alternative statement of the constancy of
the velocity of light:

If the time T and distance D between two events are related by $D = cT$
in one frame of reference, then they will be related in the same way in any
other frame of reference. Putting it another way, if the time between two

events in nanoseconds is equal to the distance between them in feet in any one frame of reference, then the time between them in nanoseconds in any other frame of reference will be equal to the distance between them in feet in that other frame.

It is sometimes useful to define T or D to be positive or negative depending on conventions about the time order of the events or the direction from one event to the other. Since two quantities that differ only in their signs have the same squares, we can include all these possibilities in our alternative statement of the constancy of the velocity of light by writing the relation in terms of the squares of the distance and time between the events, asserting that if the time and distance between two events in one frame of reference satisfy $(cT)^2 = D^2$, then they will satisfy this same condition in any other frame of reference. Equivalently, if the time and distance between a pair of events satisfy

$$c^2 T^2 - D^2 = 0 \qquad (8.1)$$

in one frame of reference, then they must satisfy (8.1) in any other frame of reference.

Two events that are separated by a time and a distance satisfying (8.1) are said to be *light-like separated*. The term is intended to remind you that a single photon can be present at both events—i.e. a photon can be produced at the earlier event that arrives at the later event just as the later event is taking place. Two such events can be bridged by a light signal. Using this terminology we can give an alternative statement of the constancy of the velocity of light: if two events are light-like separated in one frame of reference, they will be light-like separated in all frames of reference.

When stated in this way, the principle of the constancy of the velocity of light is a special case of a much more general rule. We show below that if T is the time and D is the distance between *any* two events E_1 and E_2 in a particular frame of reference, then even when $c^2 T^2 - D^2$ is not zero, its value is still the same in all frames of reference, although T and D separately vary from one frame to another. This is called the *invariance of the interval*:

> For any pair of events a time T and a distance D apart, the value of $c^2 T^2 - D^2$ does not depend on the frame of reference in which T and D are specified.

To see why this is so, we take T and D to be the (positive) magnitude of the time and distance between the events and consider separately the two different ways in which $c^2 T^2 - D^2$ can be nonzero: either cT is bigger than D, or cT is less.

Suppose first that cT is bigger than D. Then D/T is less than the speed of light c, so there is a frame of reference, moving from the earlier event to the later one at a speed

$$v = D/T, \tag{8.2}$$

in which both events happen in the same place. Let T_0 be the time between the two events in this special frame. A clock that is present at both events is stationary in the special frame, and will therefore advance by T_0 between the events.

This is illustrated in figure 8.1 from the point of view of the original frame, in which the events are separated in space and time by D and T. Because the clock moves with speed v in the original frame, the amount T_0 it advances between the events is reduced from the time T between the events by

$$T_0 = sT = T\sqrt{1 - v^2/c^2}. \tag{8.3}$$

Since $v = D/T$, it follows from (8.3) that

$$T_0^2 = T^2 - D^2/c^2. \tag{8.4}$$

So when the time T and distance D between two events are related by $T > D/c$, then $T^2 - D^2/c^2$ is independent of the frame of reference in which D and T are evaluated, being equal to the square of the time T_0 between the two events in the frame in which they happen at the same place.

The other case, $cT < D$, is a bit more subtle. Now D/T exceeds c, so it is impossible for anything moving at less than the speed of light to be present at both events. But now there is a frame of reference, moving at less than the speed of light, in which the two events happen *at the same time.* To see why, consider two clocks that are stationary and synchronized in the frame in which the events are separated in space and time by D and T, with one clock present at each event. This is illustrated in figure 8.2. If the clock at the earlier event reads 0, then the clock at the later one reads T. Since the distance between the clocks is D, we can arrange for the stationary clocks to be attached to the ends of a stationary stick of proper length D.

In a new frame, moving with speed v along the stick in the direction from the earlier event to the later one, the clock at the earlier event is behind the clock at the later one by Dv/c^2, so if we could pick v so that Dv/c^2 were equal to T, then the two events would be simultaneous in the

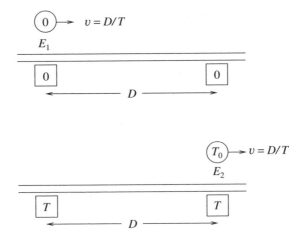

Figure 8.1. The two parts of the figure show events E_1 and E_2, taking place a distance D and a time T apart in the track frame. In the upper part of the figure two track-frame clocks (square boxes pictured just below the tracks) read 0. In the lower part of the figure the two track-frame clocks read T. A third clock (the round object above the tracks) moves with speed $v = D/T$ and is therefore able to be present both at E_1 in the upper part of the figure and at E_2 in the lower part.

new frame. This requires that

$$v = c^2 T/D = \left(\frac{cT}{D}\right) c. \tag{8.5}$$

Since D is larger than cT, the required speed v is less than c, and therefore there is indeed a frame in which the two events are simultaneous: the frame of the rocket in figure 8.2.

In the rocket frame the events occur at opposite ends of a stick of proper length D that is moving with speed $v = c^2 T/D$. Since the events are simultaneous in the rocket frame, the stick does not move between events, and the distance D_0 between the events is just the shrunken length of the moving stick:

$$D_0 = sD = D\sqrt{1 - v^2/c^2}. \tag{8.6}$$

Since the speed v of the rocket frame is given by (8.5), we deduce from (8.6) that

$$D_0^2 = D^2 - c^2 T^2. \tag{8.7}$$

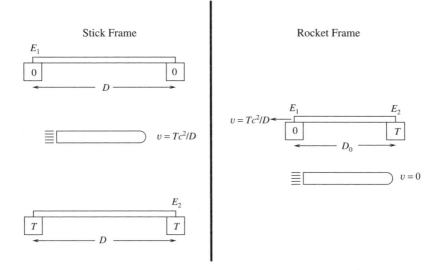

Figure 8.2. The left half of the figure shows events E_1 and E_2 that occur a time T and distance D apart in the frame of a stick of length D, at opposite ends of which the events take place. The upper left part shows E_1 occurring at time 0, and the lower left part shows E_2 occurring at time T, all times being indicated by clocks attached to the two ends of the stick and synchronized in the stick frame (square boxes pictured just below the stick). A rocket (the long object in the middle of the left half of the figure) moves to the right with speed $v = Tc^2/D = c(cT/D)$, which is less than c when $D > cT$. In the rocket frame (right half of the figure) the stick and attached clocks move to the left with speed v. In the rocket frame, when the clock on the left reads 0, the clock on the right reads $vD/c^2 = T$, so the events E_1 and E_2 are simultaneous. Therefore the distance D_0 between the events in the rocket frame is just the shrunken length sD of the moving stick.

So when the time T and distance D between two events are related by $D/c > T$, then $D^2 - c^2T^2$ is independent of the frame of reference in which D and T are evaluated, being equal to the square of the distance D_0 between the two events in the frame in which they happen at the same time.

Note the pleasing resemblance between this italicized conclusion and the italicized conclusion immediately following (8.4). When distances and times are measured in feet and nanoseconds (so that $c = 1$), the two statements differ only by the interchange of space and time.

To summarize, if D is the distance and T the time between two events then the quantity $c^2T^2 - D^2$ is independent of the frame of reference in which D and T are measured, and it is useful to distinguish between three cases:

1. $c^2T^2 - D^2 > 0$. The events are said to be *time-like separated*, because there is a frame of reference in which they happen at the same place. In that frame they are separated *only* in time, and the time T_0 between them is given by $c^2T_0^2 = c^2T^2 - D^2$.
2. $c^2T^2 - D^2 < 0$. The events are said to be *space-like separated*, because there is a frame of reference in which they happen at the same time. In that frame they are separated *only* in position, and the distance D_0 between them is given by $D_0^2 = D^2 - c^2T^2$.
3. $c^2T^2 - D^2 = 0$. The events are said to be *light-like separated*, because a single photon can be present at both events.

In the first case, note that once you *know* that $c^2T^2 - D^2$ is independent of the frame in which D and T are measured, then it is obvious that $c^2T^2 - D^2$ is given by $c^2T_0^2$ since T_0 is the time between the events in the frame in which the distance D_0 between them is 0. It is also clear that in this case there can be no frame in which the events happen at the same time, since in such a frame T would be zero, and $c^2T^2 - D^2$ could not be positive. Similarly, in the second case, given that $c^2T^2 - D^2$ is indeed invariant, the value of $D^2 - c^2T^2$ is obviously D_0^2 in the frame in which the events happen at the same time, since in that frame the time T_0 between them is zero. In this case there can be no frame in which the events happen at the same place, since in such a frame D would be zero and $c^2T^2 - D^2$ could not be negative.

The quantity

$$I = \sqrt{|c^2T^2 - D^2|}$$

is called the *interval* between the two events. The word is carefully selected to be neutral as to whether the separation it suggests is in space or in time. When $c^2T^2 - D^2$ is positive, the interval I between the events (divided by c—yet another advantage of using feet and nanoseconds as the units of space and time is that it makes this parenthetical remark unnecessary) is just the time between them in the frame of reference in which they happen at the same place. When $c^2T^2 - D^2$ is negative, the interval I between the events is just the distance between them in the frame of reference in which they happen at the same time.

There is an analogy between this state of affairs and the purely spatial description of points in a plane, illustrated in figure 8.3. Suppose we have two points P_1 and P_2 and suppose that P_1 is a distance x to the *east* of P_2, and a distance y to the *north*. Then by the Pythagorean theorem, the direct distance d between the points is given by

$$d^2 = x^2 + y^2. \tag{8.8}$$

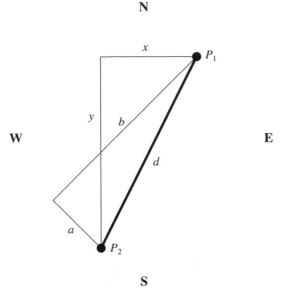

Figure 8.3. The distance d between two points P_1 and P_2 is the length of the heavy line. This is the hypotenuse of a right triangle with sides x and y, and also the hypotenuse of a second right triangle with sides a and b. The distances x and y are the east-west and north-south distances between P_1 and P_2, while the distances a and b are the northwest-southeast and northeast-southwest distances between the same two points.

If, on the other hand, P_1 is a distance a to the *northwest* of P_2 and a distance b to the *northeast*, then again by the Pythagorean theorem, the direct distance d between the points satisfies

$$d^2 = a^2 + b^2. \tag{8.9}$$

Since the direct distance between the points has nothing to do with whether you calculate it out of eastern and northern separations or northeastern and northwestern separations, we conclude that $x^2 + y^2 = a^2 + b^2$.

The remarkable discovery of relativity is that a similar relation holds for combined spatial and temporal separations. The only difference is that one subtracts rather than adds the squares to get the invariant quantity. The fact that an additional factor of c appears in the invariant quantity $c^2 T^2 - D^2$ is not a significant difference, for if we had chosen to measure eastern separation x and northeastern separation b in one set of units (say yards) and northern separation y and northwestern separation a in another (say ordinary feet), a similar conversion factor between the units would have had to appear in the purely geometrical relation (8.8). If we continued to express d in feet, then since there are 3 ordinary feet to a yard, the relation would become $d^2 = 9x^2 + y^2 = 9b^2 + a^2$. The factor c, which disappears if we use "natural units" of space and time like feet and nanoseconds, is just a conversion factor that appears if we use inappropriate units like feet and seconds ($c = 1,000,000,000$ feet per second) instead of feet and nanoseconds ($c = 1$ foot per nanosecond).

The invariance of the interval took so long to be discovered because of the enormous size of the speed of light c on the kinds of scales we are used to using. For temporal and spatial separations familiar under everyday terrestrial conditions, T is simply not small enough, nor D large enough, for cT not to be enormously larger than D, so that $I^2 = c^2 T^2 - D^2$ is hardly distinguishable from $c^2 T^2$. Under these circumstances, the invariance of the interval reduces to the assertion that the time between any pair of events is the same in all frames of reference, which is exactly what people used to believe. Only when D becomes so large and/or T so small that D/T is no longer tiny compared with c does the invariance of the interval have the richer implications we now know it to have.

Here is an entertaining consequence of the invariance of the interval. Consider two events in the history of a uniformly moving clock, a time T and a distance D apart. Since the distance between the two events is $D_0 = 0$ in the proper frame of the clock, the time T_0 ticked off by the clock between the events satisfies $T_0^2 = T^2 - D^2/c^2$, as we have already noted in (8.4). We can rewrite this relation in the form $T_0^2 + D^2/c^2 = T^2$ or, dividing both sides by T^2, as

$$T_0^2/T^2 + D^2/c^2 T^2 = 1. \qquad (8.10)$$

Since the clock is present at both events, D/T is just the speed v of the clock in the frame in which it moves; it tells us how many feet the position of the clock changes per nanosecond of time. On the other hand, T_0/T tells us how many nanoseconds the clock ticks off per nanosecond of time in the frame in which it moves. So (continuing to use feet and nanoseconds) we have

$$(T_0/T)^2 + v^2 = 1. \qquad (8.11)$$

The relation (8.11) tells us that the sum of the square of the speed at which a uniformly moving clock runs (in nanoseconds of clock reading per nanosecond of time) plus the square of the speed at which the clock moves through space (in feet of space per nanosecond of time) is 1. (This fact can also be deduced directly from the form of the slowing-down factor $s = \sqrt{1 - v^2/c^2}$ for moving clocks, without exploiting the concept of interval.)

Now a stationary clock moves through time at 1 nanosecond per nanosecond and does not move through space at all. But if the clock moves, there is a trade-off: the faster it moves through space—i.e. the larger v is—the slower it moves through time—i.e. the smaller T_0/T is—in such a way as to maintain the sum of the squares of the two at 1. It is as if the clock is always moving through a union of space and time—spacetime—at the speed of light. If the clock is stationary then the motion is entirely through time (at a speed of 1 nanosecond per nanosecond). But in order to move through space as well, the clock must sacrifice some of its speed through time, in order to keep the total speed through spacetime equal to 1, as required by (8.11).

The analogy with ordinary speed along a highway is striking: a car moving east with its cruise control set to a fixed speed of 80 feet per second must sacrifice part of its easterly speed v_e to acquire some northerly speed v_n, because the cruise control keeps the speed of the car fixed at 80 feet per second, while the Pythagorean theorem requires the easterly and northerly speeds to be related by $80^2 = v_e^2 + v_n^2$.

It is possible to measure the interval between any two events *using a single clock* that is present at only one of them. Suppose Alice is at event E_1 and Bob is at event E_2. Suppose each of them is able to see (with a telescope, if necessary) what is going on in the vicinity of the other. If Alice moves uniformly and carries a clock, then she and Bob can measure the interval between the two events as follows:

(1) Alice notes the reading t_1 of her clock when the event E_1 takes place next to her. (2) Alice notes the reading t_2 of her clock when she sees (through her telescope) event E_2 taking place next to Bob. (3) Bob notes the time t_3 that he sees (through his own telescope) on Alice's clock at the moment the event E_2 takes place next to him. The squared interval between E_1 and E_2 is given by

$$I^2 = c^2 |(t_2 - t_1)(t_3 - t_1)|. \tag{8.12}$$

If you think about it for a minute, this is clearly correct for light-like separated events for then a photon can travel from one event to the other so either $t_3 = t_1$ or $t_2 = t_1$, depending on whether the photon travels from E_1 to E_2, or the other way around. In either case (8.12) gives $I = 0$. It is also correct if Alice's clock is at E_2 as well as at E_1, which could happen if the events were time-like separated. One way to confirm (8.12) more generally is to show that if there is a frame in which the events happen at the same time (space-like separated events), then in that frame (8.12) does indeed give the square of the distance between them. And if there is a frame in which the events happen at the same place (time-like separated events), then in that frame it gives the square of the time between them. In each of these cases there are two subcases: for space-like separated events Alice can move toward or away from E_2; for time-like separated events E_1 can occur before or after E_2. In all of these cases, one takes advantage of the fact that time differences can be inferred by dividing differences in readings of Alice's clock by the slowing-down factor s_v determined by her velocity.

You might examine a few of these cases. I shall postpone my own proof of (8.12) to chapter 10, pp. 138–40, where we can verify it by a much more powerful and intuitive procedure that simply draws a few pictures.

Nine

Trains of Rockets

IN THIS CHAPTER WE SHALL EXAMINE an easy way to explore how a dis-agreement about whose clocks are synchronized leads to all the relativistic effects we have found: the slowing down of moving clocks, the shrinking of moving sticks, the relativistic velocity addition law, the existence of an invariant velocity, and the invariance of the interval.

We shall do this by examining two frames of reference from the point of view of a third frame in which the first two move with the same speed, but in opposite directions. We take the third frame to be the proper frame of a space station. The first two frames are the proper frames of two trains of rockets: a gray train, moving to the left in the frame of the space station, and a white train, moving to the right in the frame of the space station, at the same speed that the gray train moves to the left.

Figure 9.1 shows the station (represented by a black circle) and the two trains of consecutively numbered rockets at four different moments of time, as described in the frame of the station. The station is in the same place in all four parts of the figure, while each train has moved an additional rocket's length from one part of the figure to the next. The three numbers preceded by a colon (e.g., :006) adjacent to each rocket represent the reading of a clock carried by that rocket. Each clock is at the center of its rocket, right next to the number of the rocket.

Notice that the clocks on either train of rockets, in each of the four parts of the figure, are not synchronized: as you go toward the rear of either train, the clocks get further and further ahead, the asynchronization being exactly two temporal units, which we shall call "ticks," per rocket of separation. This is in accord with the station-frame rule that if clocks have been synchronized in the train frame, then they are out of synchronization in the station frame, a clock in front being behind a clock in the rear by $T = Du/c^2$, where D is the distance between the clocks in their proper frame, and u is the speed of the train in the station frame. If we take our unit of length to be the proper length of a rocket, then evidently the figure has been drawn for a value of u for which the asynchronization is

$$u/c^2 = 2 \text{ ticks per rocket.} \tag{9.1}$$

You can take two attitudes toward figure 9.1. You can imagine that both trains are moving with such prodigious speeds and the tick is such a tiny

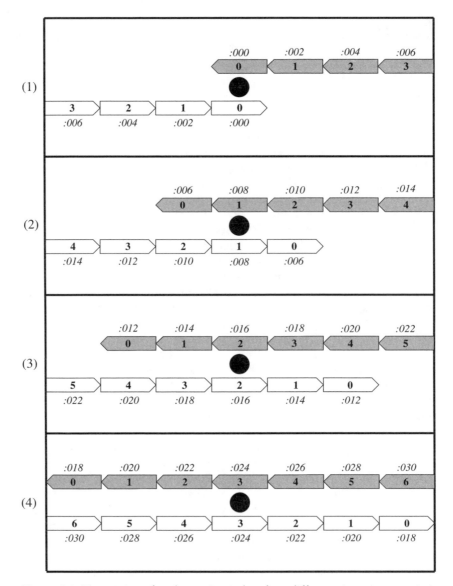

Figure 9.1. Two trains of rockets, pictured at four different times in parts 1–4. Numbers (like :014) above or below a rocket represent the reading of a clock attached to that rocket. Within each train the clocks on adjacent rockets are out of synchronization by :002. Between successive pictures the reading of each clock advances by :006, and each train moves one rocket's length to the left (gray train) or right (white train). The black circle represents a space station.

unit of time and the clocks are so very precise that the asynchronization depicted in the figure is the genuine relativistic effect: u/c^2 nanoseconds of asynchronization for every foot of separation.

Alternatively, and more entertainingly, one can take the view that the rockets are moving at perfectly feasible speeds—perhaps a few feet per millisecond—and the clocks are quite ordinary clocks, ticking off seconds with good but not phenomenal precision, which have, however, been deliberately set out of synchronization by people in the space station. Under this reading of the figure, the station people want to test what kind of conclusions people on either train will arrive at using unsynchronized clocks, if they fail to realize that their clocks are not synchronized. So before the trains start to move, but after people have been locked into their rockets, the station people give the occupants of each rocket a clock. But they secretly set the clocks ahead by two ticks per rocket as they move from the front toward the rear of the train, distributing the clocks. They also carefully arrange things so that people in different rockets cannot communicate with people in other rockets of their train to compare notes on what their clocks read. The space station people then lie to the occupants of each train, falsely assuring them that clocks in different rockets are synchronized. Let us pursue this alternative conspiratorial interpretation.

Once the trains are set into motion, people from either train can collect information only about what is going on in their immediate vicinity. In particular when two rockets are directly opposite each other (for example, in part 1 white and gray rockets 0 are directly opposite, in part 3 gray rocket 1 is directly opposite white rocket 3), then the occupants of either rocket can note the number and clock reading of the other rocket, as well, of course, as the number and clock reading of their own rocket. Such information can be summarized in a little figure that shows just those two adjacent rockets. Figure 9.2, for example, is a fragment of part 4 of figure 9.1, showing information the occupants of gray rocket 1 or white rocket 5 might have acquired at the moment they were directly opposite.

Contemplating this picture, inhabitants of the white train would say that at a white time of 28 ticks gray rocket 1 was opposite white rocket 5 and its clock read 20 ticks. Inhabitants of the gray train, looking at the same picture, would say (equivalently) that at a gray time of 20 ticks white rocket 5 was opposite gray rocket 1 and its clock read 28 ticks. Note that the only difference in interpretation of the picture is that inhabitants of each train regard their clock as telling the correct time, and the clock on the other train as incorrectly set, with a reading that does not give the actual time at which the picture was taken.

Suppose after the trains have gone past one another and large quantities of such information have been collected by the occupants of both trains,

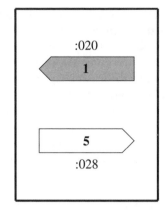

Figure 9.2

they return to the space station. Then the people from each train go off to two separate conference rooms—one for the gray train and one for the white train—to compare notes on what pictures they took. What conclusions can they draw, if they act under the assumption that all the different clocks on their own train are synchronized?

The first interesting thing to examine is any pair of pictures in which the same rocket appears. Figure 9.3, for example, shows two pictures containing grey rocket 0, taken from parts 2 and 3 of figure 9.1. People on the white train will interpret these pictures as follows:

One thing they can read off from the two pictures is the speed of gray rocket 0, for in the first pictures it is opposite white rocket 2 at a time of 10 ticks, while in the second picture it is opposite white rocket 4 at a time of 20 ticks. So gray rocket 0 took 10 ticks to go two rockets and is therefore traveling at a speed of $\frac{1}{5}$ rocket per tick.

Another thing the people on the white train can note from the same two pictures is that at the white time of 10 ticks the clock on gray rocket 0 reads 6 ticks, while at the later white time of 20 ticks the clock on gray rocket 0 reads 12 ticks. Therefore in the actual white time of 10 ticks that elapsed between the taking of the two pictures, the gray clock only advanced by 6 ticks. So it is running slowly by a factor of $\frac{3}{5}$.

Note that the validity of both these conclusions depends crucially on the assumption that the white clocks are synchronized, since the people from the white train are using the readings of two *different* clocks (one in white rocket 2 and the other in white rocket 4) to make their judgments about the times at which events take place. You can check for yourself that any other pair of pictures containing a single gray rocket leads to the same conclusion according to the white train: the speed of that rocket is $\frac{1}{5}$ rocket per tick, and its clock is running slowly by a factor of $\frac{3}{5}$.

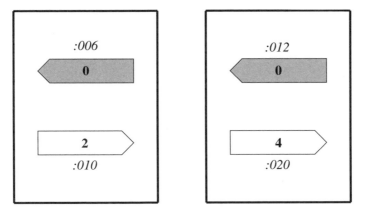

Figure 9.3

Since figure 9.1 is symmetric between gray and white, the gray-train people, as you can easily confirm, will reach exactly the same conclusion about the white train and its clocks from any pair of pictures that show a single white rocket. They will conclude that the white train is moving at $\frac{1}{5}$ rocket per tick and all its clocks are running slowly by the factor $\frac{3}{5}$. This directly and straightforwardly reveals how a disagreement about whose clocks are correctly synchronized can lead to occupants of each of the two trains being firmly convinced that it is the clocks on the other train that are running slowly. We, of course, taking the view appropriate to the station frame in figure 9.1, believe that both sets of clocks are running at exactly the same rate, and that neither set of clocks is correctly synchronized.

We now have both the speed, $v = \frac{1}{5}$ rocket per tick, that the occupants of one train assign to the other train, as well as the slowing-down factor, $s = \frac{3}{5}$, that they assign to the clocks on the other train. Anticipating that these ridiculously simple pairs of pictures extracted from the ridiculously simple full set of pictures in figure 9.1 are going to mimic all of the relativistic effects, we can note that an s of $\frac{3}{5}$ is associated with v/c of $\frac{4}{5}$, since $s = \sqrt{1 - v^2/c^2}$. So with $v = \frac{1}{5}$ rocket per tick, we have $\frac{v}{c} = \frac{4}{5} = \frac{1/5}{1/4}$, and we should therefore be on the alert in what follows for the speed of $\frac{1}{4}$ rocket per tick to play the role of an invariant velocity—the speed of light.

The next interesting thing we can do is to examine any pair of pictures that were taken at the same time, according to one of the trains. Consider, for example, the two pictures taken at the gray time of 20 ticks, extracted from parts 3 and 4 of figure 9.1, and shown in figure 9.4. Because these pictures were taken at the same time, according to the occupants of the gray train, they immediately reveal to the gray-train people that the clocks

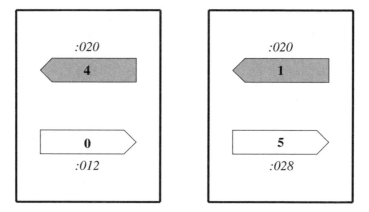

Figure 9.4

on the white train are not synchronized. For at the gray time of 20 ticks the clock in white rocket 0 reads 12 ticks, but that in white rocket 5 reads 28 ticks. The white clocks disagree by 16 ticks and are five white rockets apart, so they are out of synchronization by $\frac{16}{5} = 3.2$ ticks per rocket.

You should not be surprised that this is different from the asynchronization of exactly 2 ticks per rocket evident in figure 9.1. A difference is to be expected, since figure 9.1 depicts things as they are in the station frame, in which the gray clocks are as badly out of synchronization as the white ones, and therefore just as unreliable.

People on the gray train can also conclude from figure 9.4 that at a single moment of gray time—20 ticks—five white rockets (rockets 4, 3, 2, 1, and half each of rockets 5 and 0) stretched the same length as three gray rockets (rockets 2, 3, and half each of rockets 4 and 1), so the white rockets have shrunk by the same factor of $\frac{3}{5}$ as the white clocks are running slowly.

The same conclusions can be reached from any other pair of pictures in which the gray clocks read the same, and the same conclusions except for the interchange of gray and white can be reached from any pair in which the white clocks read the same.

Notice that a clock asynchronization of 3.2 ticks per rocket is just what you would expect from the $T = Dv/c^2$ rule with the values we have found for v and c. For with $v = \frac{1}{5}$ rocket per tick and $c = \frac{1}{4}$ rocket per tick, v/c^2 is

$$\frac{\frac{1}{5}}{(\frac{1}{4})^2} = \frac{16}{5} = 3.2 \text{ ticks per rocket,} \tag{9.2}$$

as we read off directly from figure 9.4.

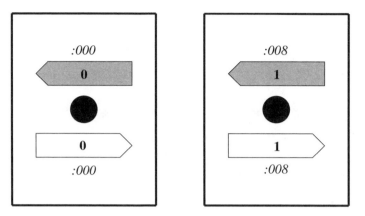

Figure 9.5

We can also reconcile all this information with the fact, evident from figure 9.1, that people using the station frame know that the clock asynchronization on *both* trains is 2 ticks per rocket. If u is the speed of either train in the station frame, the clock-asynchronization rule tells us that u/c^2 should be 2 ticks per rocket. If the invariant velocity is indeed $c = \frac{1}{4}$ rocket per tick, then u, the speed of either train in the station frame, ought to be $\frac{1}{8}$ rocket per tick, since $2 = \frac{1/8}{(1/4)^2}$. We can verify that this is indeed the case by using the fact that the speed of a train in the station frame is the same as the speed of the station in the train frame. It is evident from the pictures from parts 1 and 2 of figure 9.1, reproduced in figure 9.5, that according to either train the station moves from rocket 0 to rocket 1 in a time of 8 ticks, so its speed is indeed $\frac{1}{8}$ rocket per tick.

We can also check that these various speeds are consistent with the relativistic velocity addition law,

$$v_{wg} = \frac{v_{ws} + v_{sg}}{1 + v_{ws}v_{sg}/c^2}, \tag{9.3}$$

where v_{wg} is the velocity of the white train in the frame of the gray train, v_{ws} is the velocity of the white train in the station frame, and v_{sg} is the velocity of the station in the frame of the gray train. We know that $v_{ws} = v_{sg} = \frac{1}{8}$ rocket per tick and $c = \frac{1}{4}$ rocket per tick. When these numbers are put into (9.3), the result is indeed, $v_{wg} = \frac{1}{5}$ rocket per tick, so the precise quantitative relativistic relations continue to hold.

I emphasize how very little has gone into the construction of figure 9.1. The structure of part 1 of that figure is extremely simple. The only peculiar thing about it is the fact that the clocks do not all agree with each other. The way in which they disagree is evident. The rule for getting each of the

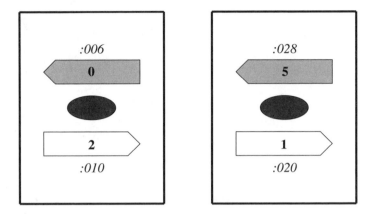

Figure 9.6

other parts of the figure from the part above it is just to shift each train by one rocket in the direction in which it is going, and advance every clock on each train by 6 ticks. Nothing elaborate has to be done to get relativity out of the figures. Once one introduces the asynchronized clocks on each train, all the other relativistic effects follow automatically.

The relativistic velocity addition law works for *anything* that moves along the two trains—not just the station itself. Consider, for example, an object that was between gray rocket 0 and white rocket 2 in part 2 of figure 9.1 and between gray rocket 5 and white rocket 1 in part 4. It has been captured in the two pictures shown in figure 9.6. According to the gray train the object has gone five rockets to the right in 22 ticks, and according to the white train it has gone one rocket to the right in 10 ticks, so we have $v_{og} = \frac{5}{22}$ rocket per tick and $v_{ow} = \frac{1}{10}$ rocket per tick. We should have

$$v_{og} = \frac{v_{ow} + v_{wg}}{1 + v_{ow}v_{wg}/c^2}, \tag{9.4}$$

which gives

$$v_{og} = \frac{\frac{1}{10} + \frac{1}{5}}{1 + (\frac{1}{10})(\frac{1}{5})/(\frac{1}{4})^2}, \tag{9.5}$$

which does indeed give $v_{og} = \frac{5}{22}$ rocket per tick after all the arithmetic is carried out.

You can check for yourself that any other pair of pictures extracted from figure 9.1 containing two moments in the history of a single object yields values of v_{ow} and v_{og} that are consistent with the relativistic velocity

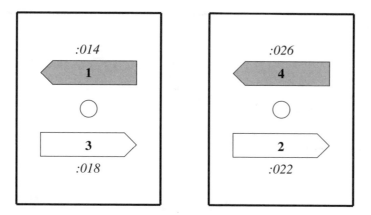

Figure 9.7

addition law (9.4) and the facts that $v_{wg} = \frac{1}{5}$ rocket per tick and $c = \frac{1}{4}$ rocket per tick. (But be sure to give v_{ow} and v_{og} the right signs: positive if the object is moving toward the right end of the train, and negative if it is moving toward the left end.)

It is particularly instructive to hunt around for a pair of photographs displaying two moments in the history of a single object moving at the special speed of $\frac{1}{4}$ rocket per tick. Figure 9.7 shows such a pair, taken from parts 3 and 4 of figure 9.1. I have added an object present in both figures, in the form of a white circle. According to the gray train, the object has moved three rockets to the right in a time of 12 ticks, so its velocity is $\frac{1}{4}$ rocket per tick. And according to the white train, it has moved one rocket to the right in a time of 4 ticks, so its velocity is again $\frac{1}{4}$ rocket per tick.

Such an object has the entertaining ability to exploit the differences in clock synchronization on the two trains, in such a way that it can move along either train in the same direction and at the same speed, $\frac{1}{4}$ rocket per tick, provided, of course, that the speed along a given train is timed by using the clocks carried by the rockets in that train, and provided those clocks are assumed to be synchronized. In figure 9.8 I have drawn another such object (moving to the left in the frame of each train) to each of the four parts of figure 9.1, so you can have a broader view of the elegant way it manages to move along both trains so that occupants of either train assign to it a speed of $\frac{1}{4}$ rocket per tick.

Figure 9.1 also provides us with a new insight into why motion faster than light is highly problematic. Figure 9.9 shows two pictures in the history of a hypothetical faster-than-light object taken from parts 3 and 4 of figure 9.1. I have added to each picture a black oval, representing the object. According to the gray train, it has gone six rockets in 18 ticks, for a speed of $\frac{1}{3}$ rocket per tick, which exceeds the invariant velocity $c = \frac{1}{4}$

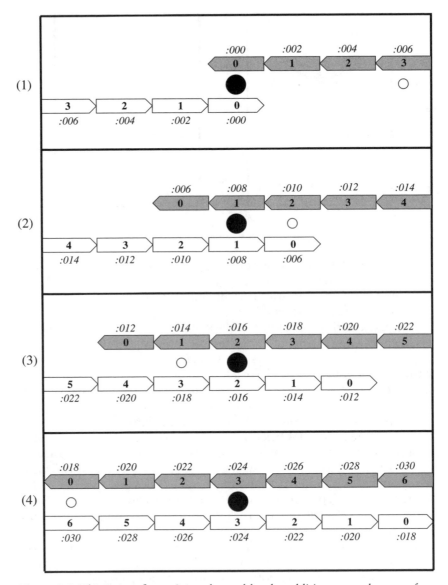

Figure 9.8. This is just figure 9.1, enhanced by the addition, to each part, of an object (the small white circle) whose speed, according to either train, is $\frac{1}{4}$ rocket per tick.

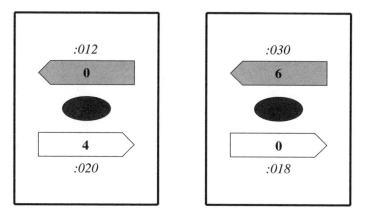

Figure 9.9

rocket per tick. People from the white train agree that the object goes faster than the invariant velocity, having gone four rockets in a mere 2 ticks, for a speed of 2 rockets per tick. (You can check that even these superluminal velocities are consistent with the relativistic velocity addition law. But be careful with the signs that indicate which way the object is moving in the frame of each train.)

But there is a disturbing aspect to figure 9.9. According to the gray train, the picture on the left was taken 18 ticks (:030−:012) *before* the one on the right. On the other hand, according to the white train, the picture on the left was taken 2 ticks (:020 − :018) *after* the one on the right. Occupants of the two trains disagree about the order in which the two pictures were taken! This is the kind of disagreement it is hard to tolerate. Suppose, for example, that the object were a burning candle. Its pictures would then clearly reveal the direction of time: the later the picture, the shorter the candle and the bigger the puddle of wax beneath it. Such a pair of pictures would reveal to one of the two groups that the clocks on its own train could not have been telling the correct time.

It turns out that this situation is quite general. If an object moves faster than light then there are always two frames of reference that disagree about the order in time of any pair of events in the history of the object. This is most easily demonstrated using the space-time diagrams of chapter 10, in figure 10.18. If anything could move faster than light, it would have to be incapable of revealing, through its internal structure, any information about the direction of the flow of time. Burning candles, melting ice-cubes, rotting bananas, running-down batteries, aging people, and the like cannot move faster than light.

Another anomalous feature of faster-than-light motion, noted at the end of chapter 5, is also present in figure 9.1. Anything that moves at a

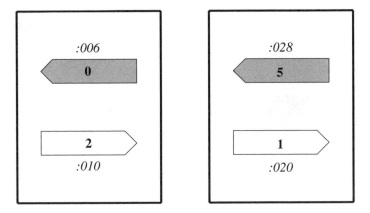

Figure 9.10

speed of $\frac{5}{16}$ rocket per tick (which exceeds $\frac{1}{4}$ rocket per tick) according to one train will be present in all rockets at one and the same time according to the other. I invite you to confirm this by examining the figure.

Note, finally, that figure 9.1 can be used to demonstrate the invariance of the interval between two events. Take any two little pictures whatever and calculate

$$T^2 - D^2/c^2 = T^2 - (4D)^2 \tag{9.6}$$

(the 4 comes from the c^2, since c is $\frac{1}{4}$ rocket per tick), where T is the number of ticks between the events in the two pictures and D, the number of rockets separating the events. The answers will not depend on which frame you take to evaluate T and D. Consider, for example, figure 9.10, which takes one event from part 2 and another from part 4 of figure 9.1. According to the gray frame, the two events are 22 ticks and five rockets apart, and $22^2 - (4 \times 5)^2 = 22^2 - 20^2 = 84$. According to the white frame, the two events are 10 ticks and one rocket apart, and $10^2 - (4 \times 1)^2 = 10^2 - 4^2 = 84$. This particular pair of events is time-like separated, since $T^2 - D^2/c^2$ is positive, and, indeed, an object present at both events would have a speed less than $\frac{1}{4}$ rocket per tick in either frame ($\frac{5}{22}$ rocket per tick in the gray frame and $\frac{1}{10}$ rocket per tick in the white frame). By picking other pairs of pictures randomly from figure 9.1, you can convince yourself that $T^2 - (4D)^2$ always comes out the same for the time and distances between two events associated with the two pictures, whether you read off T and D from the white rockets or the gray ones.

I have described all this as if the clocks on both trains were deliberately set out of synchronization by the people in the station frame. If that is how the trains and clocks were really set up, and if the people on either train were under the impression that their clocks were actually synchronized, then they would interpret their photographs exactly as we have done.

What is special about the world we live in is this: if the people in the station frame should choose to do the experiment for trains moving with a speed of u feet per nanosecond, and should they choose the clock asynchronization to be exactly u nanoseconds of disagreement per foot of rocket, then to set up the clocks on both trains all they would have to do would be to furnish people on each train with a highly accurate set of clocks, set the trains into motion, and instruct the people on each train to synchronize their clocks. Nature herself would automatically provide the discrepancy between the station-frame interpretation of the clocks and the interpretation from within each train.

Ten

Space-Time Geometry

IN VARIOUS FIGURES THROUGHOUT THE BOOK, we have examined different events—things happening at a definite place at a definite time—that occur along a straight railroad track or along a straight line of rockets. Examples of such an event are two rockets and their attached clocks being directly opposite one another, or a signal arriving at the end of a train and triggering the making of a mark on the tracks. In the figures, these events are represented by regions that are small on the scale of the entire figure. In most of the figures, we have taken a horizontal separation of two such regions to indicate, in an appropriate frame of reference, a spatial separation of the events they represent, and a vertical separation of different parts of a figure to indicate a temporal separation of the events they contain.

In this chapter we shall develop a somewhat more abstract and substantially more powerful generalization of such figures. In these new figures the regions representing individual events are collapsed to geometric points. By exploring this graphical way of representing phenomena a little more generally and systematically, it is possible to arrive at a deeper—I would say, in fact, the deepest—understanding of what relativity has to tell us about the nature of space and time. The figures we shall be constructing are called *space-time diagrams* or *Minkowski diagrams*, after Hermann Minkowski, who invented them in 1908, a mere three years after Einstein's first relativity paper.

For simplicity we continue to deal with only one spatial dimension—all the events we shall consider take place along a single straight track. Adding the other two dimensions—horizontal and vertical distance away from the track—can sometimes give further insight, but it makes it impossible to draw everything on a page. Many important phenomena, including all those we have examined up to now, do indeed involve relative motion in only a single spatial dimension.

We start with a particular frame of reference (Alice's) and specify some simple rules that Alice can use to represent events by points on a page. Before Bob appears on the scene, everything that follows refers to Alice's frame of reference. So until then, when I talk about events happening in the same place or at the same time, I shall mean at the same place or at the same time according to Alice.

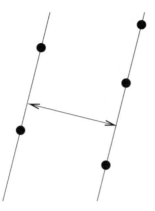

Figure 10.1. Two equilocs for Alice's frame. The two black dots on the line on the left represent two events that happen in a single place (but at different times) according to Alice; the three black dots on the line on the right represent three other events that happen in a single place, different from the place of the two events on the left. The distance in the diagram between the two equilocs (indicated by the double-headed arrow) is proportional to the actual distance in Alice's frame between the two places the equilocs represent. Such a diagram is characterized by a scale factor λ, which specifies, for example, the number of centimeters on the page between equilocs that represent places a foot apart in space.

Alice represents an event by a single point in her diagram. Two or more events that happen at the same place *and* at the same time (space-time coincidences) are represented by one and the same point, so a single point can represent either a single event or several coincident events. Distinct points in the diagram are associated with distinct events that happen either at different places, at different times, or at both different places and different times.

Alice represents several events that happen at the same place, but not at the same time, by points on a single straight line. This is illustrated in figure 10.1. We call such a line a line of constant position or, more compactly, an *equiloc*. (The term "equiloc" is not currently part of the standard language of relativity, but it ought to be.) Alice is clearly free to orient one such equiloc in any direction she chooses, since such a choice amounts to nothing more than picking an orientation for the page on which she draws her diagram.

Notice that any two such equilocs, representing various events that happen in two *different* places, have to be parallel. For if they were not, then since they are straight lines they would have to intersect somewhere. Their point of intersection would then correspond to a single event that happened in two different places. But by definition an event is something that happens at a single place (and at a single time), so this makes no sense. Distinct equilocs must be parallel.

Following the usual conventions of map makers (more precisely, those making maps of regions very small compared with the radius of the Earth), Alice takes the distance on the page between two distinct equilocs to be proportional to the actual distance between the positions of the events that they represent: the farther apart the equilocs are on the page, the farther apart are the two places.

The quantitative connection between distances in space and distances in the diagram is given by a scale factor λ. Multiplication by λ converts the actual spatial distance between two events into the distance on the page between the equilocs on which the events lie in the diagram. For example if equilocs separated by 1 centimeter on the page corresponded to events at positions 1 kilometer apart, then λ would be 1 centimeter to the kilometer, numerically 1/100,000. If we wish to distinguish Alice's scale factor from those used by other people (and it turns out to be important to be able to do this), we can give it a subscript, calling it λ_A.

These rules are, I hope, commonplace and boring. Things get no more exciting with the next few rules, which merely specify for location in time what we have just specified for location in space. Alice positions events along her equilocs so that events happening at the same time (but not necessarily in the same place) are represented in her diagram by points on a single straight line, as illustrated in figure 10.2. Such a line is called a line of constant time, or *equitemp*. (Like "equiloc," the term "equitemp" has not yet entered the standard lexicon of relativity.) Alice is free to orient one such equitemp to make any nonzero angle she wishes with her equilocs, since such a choice of direction amounts to nothing more than an appropriate stretching of the page on which she draws her diagram (considered, for this purpose only, to be made of rubber). Remembering figures from earlier chapters, you might be tempted to make Alice's equitemps horizontal, but resist the temptation. It is far too restrictive a choice, as we shall soon see.

Like equilocs, any two different equitemps must be parallel, for, being straight lines, if they were not parallel they would intersect somewhere and their point of intersection would correspond to a single event happening at two different times. But by definition an event is something that happens at a single time (and at a single place). Following her procedure for equilocs, Alice takes the distance on the page between two distinct equitemps to be proportional to the actual time interval between the times of the events they represent.

So now we have equilocs, whose points represent events that happen in a single place, and equitemps, whose points represent events that happen at a single time. Note that any equitemp can intersect any equiloc in one and

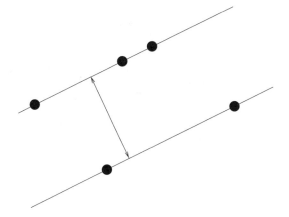

Figure 10.2. Two equitemps for Alice's frame. The two black dots on the lower line represent two events that happen at a single time (but in different places) according to Alice; the three black dots on the upper line represent three other events that happen at a single time, different from the time of the two on the lower line. The distance between two such equitemps in the diagram (indicated by the double-headed arrow) is proportional to the actual time in Alice's frame between the two moments of time the equitemps represent. Such a diagram is characterized by a scale factor λ, which specifies, for example, the number of centimeters on the page between equitemps that represent events a nanosecond apart in time.

only one point. That point represents those events that happen precisely at *that* time and in *that* place. Consequently the common direction of all Alice's equitemps, though it is otherwise hers to choose, must be different from the common direction of all her equilocs. Equitemps and equilocs must cross at some nonzero angle θ. If you are now tempted to make θ equal to 90 degrees, resist this temptation as well. It is also far too restrictive.

Now, for the first time, the fundamental importance of the speed of light has an influence, though a very elementary one, on the structure of Alice's diagrams. It turns out to be extremely convenient for Alice to take the distance in the diagram between two equitemps, representing events a nanosecond apart, to be exactly the same as the distance in the diagram between two equilocs, representing events a foot apart. (Recall that we have redefined a foot to be the distance light travels in vacuum in a nanosecond.) This means that the scale factor λ for equilocs (in, for example, centimeters of diagram per foot) is numerically the same as the scale factor λ for equitemps (in centimeters of diagram per nanosecond.)

With this scale convention, it follows that another convenient scale factor, the distance μ along any equiloc, associated with two events a nanosecond apart in time, is exactly the same as the distance along any

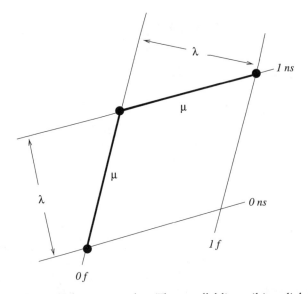

Figure 10.3. The scale factors λ and μ. The parallel lines tilting slightly upward to the right are equitemps; events represented by points on the upper equitemp happen 1 nanosecond after events represented by points on the lower one. The parallel lines tilting steeply upward to the right are equilocs; events represented by points on one equiloc happen 1 foot away from events represented by points on the other. The scale factor λ is the distance in the diagram between the equilocs or between the equitemps. The scale factor μ is the length in the diagram of the (more heavily drawn) segments of the equitemps or equilocs between the events represented by the black dots, which are 1 foot or 1 nanosecond apart.

equitemp, associated with two events a foot apart in space. This is shown in figure 10.3. It is a consequence of the elementary geometric fact that when a pair of parallel lines intersects another pair of parallel lines separated by the same distance as the first pair—in this case the distance is just λ—then the parallelogram defined by the four points of intersection has four equal sides whose common length, in this case, is just μ.

A parallelogram with four equal sides is called a rhombus. The lines connecting opposite vertices—called "diagonals"—bisect the angles at those vertices, and are perpendicular to each other. These elementary properties of rhombi turn out to be important in what follows.

Note that the scale factor μ exceeds the scale factor λ, unless Alice ignores my warning and takes her equitemps to be perpendicular to her equilocs. In that special case, μ = λ. Both scale factors are useful. Sometimes it is easiest to extract the time (or distance) between events from the distance between the equitemps (or equilocs) on which they lie, in which case λ is the relevant scale factor. But more often one wants to extract the time (or distance) between events that happen in the same place (or at the same time) from their distance apart on an equiloc (or equitemp), in which case μ is the appropriate scale factor. We shall call a rhombus like the one in figure 10.3, whose sides are segments of equilocs and equitemps representing events 1 foot and 1 nanosecond apart, a *unit rhombus*.

A particularly important collection of events, for an object small enough, on the length scale of interest, to be considered to occupy just a single point of space at any moment of time, is the set of *all* events at which the object is present. The totality of all such events is represented by a continuous line in the diagram. This line, which represents the entire history of the object, is called the *world line* or *space-time trajectory* of the object. For example an object stationary in Alice's frame of reference throughout its entire history is represented by the equiloc associated with the place the object occupies. An object moving uniformly in Alice's frame of reference is represented by a straight line that is not parallel to any equiloc, since the object is at different positions at different times. An object that is moving nonuniformly—for example, back and forth—is represented by a wiggly line.

Different objects moving uniformly with the same velocity make parallel lines. This is because in the same time, according to Alice, they cover the same distance in the same direction. Alternatively, if their space-time trajectories were not parallel they would meet in a point. That point would represent an event in which the objects were in the same place at the same time. But if their velocities are the same they would have to be in the same place at all times and their trajectories would be identical. The trajectories of objects moving with the same speed in opposite directions (and therefore with different velocities) are not, of course, parallel. Such a pair of objects can be in the same place at just a single moment. An especially important world line is the space-time trajectory of a photon, or of any other object moving at the speed of 1 foot per nanosecond. Thanks to our choice of scales, photon trajectories have some simple geometric features.

Any two events on a photon trajectory must be as many feet apart in space as they are nanoseconds apart in time. Because of the relation we imposed on the spatial and temporal scale factors, the two equilocs passing through those two events must be the same distance apart in the diagram as the two equitemps passing through those events. As a result, the photon trajectory is the diagonal of the rhombus formed by the two pairs of equidistant equilocs and equitemps, so the trajectory bisects the angles at the two vertices it connects. This is demonstrated in figure 10.4, which reveals an extremely important property of Alice's diagrams:

The angle that equilocs make with the trajectory of a photon must be the same as the angle that equitemps make with that trajectory. Putting it another way, the two photon trajectories through the point of intersection of an equiloc with an equitemp, bisect the angles formed by those

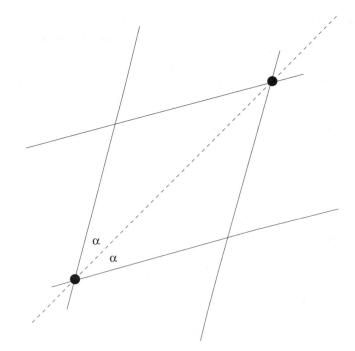

Figure 10.4. The dashed line represents the space-time trajectory of a photon. The black dots represent two events in the history of that photon. An equiloc slanting steeply upward passes through each dot, and an equitemp slanting slightly upward also passes through each. Because the photon moves 1 foot every nanosecond, the distance between the equilocs is the same as the distance between the equitemps. The parallelogram formed by the four lines is therefore a rhombus, the dashed photon line is the diagonal of that rhombus, and the symmetry of the rhombus requires the angles labeled α to be the same.

two lines. Since this rule applies to photons moving in either direction, we have a second important property, demonstrated in figure 10.5: the trajectories of two photons moving in opposite directions are *perpendicular* to each other in a diagram. This also follows directly from figure 10.4 if one notes that the other (undrawn) diagonal of the rhombus is also a photon trajectory, for a photon moving in the opposite direction, and the diagonals of a rhombus are necessarily perpendicular.

So even though Alice can freely choose the angle θ between her equitemps and equilocs, our scale convention requires certain angles to be fixed: *the world lines of oppositely moving photons are necessarily perpendicular.* This perpendicularity is a direct consequence of the relation we imposed on the spatial and temporal scale factors λ.

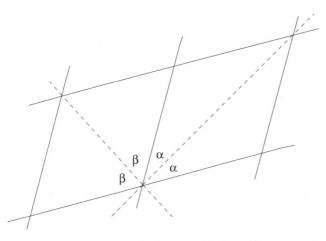

Figure 10.5. Figure 10.4 is redrawn (without the black dots) and extended to show the space-time trajectory of a second photon traveling in the opposite direction. Because the new dashed line is also a photon trajectory it also bisects the angle between the equitemps and equilocs. Since $2\alpha + 2\beta = 180$ degrees, the angle $\alpha + \beta$ between the two photon lines is 90 degrees.

Alice can rotate her page so the trajectories of two photons moving in opposite directions are symmetrically disposed about the vertical direction, tilted at 45 degrees to the right and left, with the times of the events on each photon trajectory increasing toward the top of the page. Because Alice's equitemps make the same angle with the photon trajectories as her equilocs, her equilocs will then tilt away from the vertical at the same angle that her equitemps tilt away from the horizontal. It is conventional always to orient a space-time diagram in this way, so that the vertical and horizontal directions bisect the right angles between the two families of photon trajectories, and so that equitemps lying higher in the diagram represent events occurring at later times. With this convention for orienting the diagram, equilocs are always more vertical than horizontal (i.e. their angle with the vertical is less than 45 degrees), while equitemps are always more horizontal than vertical (i.e., their angle with the horizontal is less than 45 degrees).

We have now completely determined the structure and orientation of the system of equilocs and equitemps that Alice uses to locate events in space and time, except for two choices still available to her:

1. She is free to choose the scale factor λ—i.e., the distance on the page between two equilocs associated with places 1 foot apart (which is also the distance on the page between two equitemps associated with events 1 nanosecond apart).

2. She is free to choose the angle θ that her equitemps make with her equilocs or, equivalently, the angle $\frac{1}{2}\theta$ that both families of lines make with the photon lines. (Her choice of λ and θ together fix the alternative scale factor μ.)

Alice's choice of scale depends, of course, on how big a page she has and on the spatiotemporal extent of the collection of events she wishes to represent in her diagram. Her choice of angle depends on what she (or we) wish to do with her diagram. If she is using it only for her own private purposes then a pleasing choice is to take θ to be 90 degrees, so that her equilocs are vertical, her equitemps are horizontal, and her two scale factors λ and μ are equal. If, however, she (or we) wish to compare the space-time description of events that she reads from her diagram with the space-time description of those same events provided by other observers using different frames of reference, then taking θ to be 90 degrees need not give the clearest picture. To see why, we must consider the uses to which Alice's diagrams can be put by people who prefer to describe events using other frames of reference.

Bob, moving uniformly along the track with velocity v with respect to Alice, wishes to describe the same events that she has been describing, but prefers a frame of reference in which he is the one at rest. Suppose Bob is shown Alice's diagram, filled with points that represent isolated events and lines that represent space-time trajectories, but without any of her equitemps and equilocs that she might have drawn to help her locate events in space and time. Rather than make his own independent diagram to describe all these events, Bob can use precisely the same collection of points and trajectories that Alice used. But he will describe them in a different spatiotemporal language, since he will disagree with Alice's general notions of "same place" and "same time." He will therefore not use Alice's equilocs and equitemps. It is not hard to figure out what he must do to impose his own equilocs and equitemps on Alice's diagram.

If Bob's frame of reference has velocity v with respect to Alice's, then Bob's equilocs must be parallel to the space-time trajectory of something that Alice maintains is moving with velocity v. Thus Bob's equilocs are parallel straight lines that are not parallel to Alice's equilocs. The faster Bob moves with respect to Alice, the more his equilocs tilt away from Alice's. Alice's equitemps and equilocs through any two points on one of Bob's equilocs define a parallelogram, the ratio of whose sides (or the ratio of the distances between whose sides) is just the velocity v of his frame with respect to hers in feet per nanosecond. This is illustrated in figure 10.6.

We can also determine the orientation of Bob's equitemps. This is the first place where relativity enters the story. Up until now we have made no use of either the principle of relativity or the constancy of the velocity of light, except to anticipate the important role to be played by the

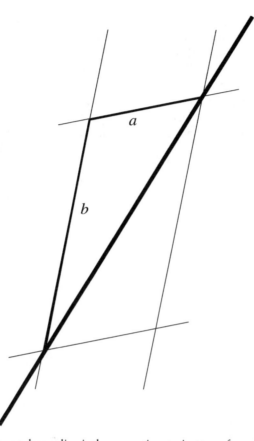

Figure 10.6. The very heavy line is the space-time trajectory of an object stationary in Bob's frame of reference—i.e. an equiloc, according to Bob. The lighter lines are equitemps and equilocs of Alice, drawn through two events on the heavy trajectory. The lengths in the diagram of the heavier segments of those lines are a and b. The velocity of Bob with respect to Alice is $v = a/b$, since in Alice's frame the position of the object changes by a distance a/μ_A in a time b/μ_A.

speed of light c by our choice of how Alice's spatial and temporal scale factors are related. Now, however, to determine the orientation of Bob's equitemps, we must put into the diagram a set of events associated with the procedure, discussed in chapter 5, that determines the simultaneity of events taking place at the two ends of a train that is stationary in Bob's frame of reference. In doing this we shall make no use of the analysis we carried out in chapter 5, but shall reach the same quantitative conclusions directly from the diagram itself, thereby demonstrating the power of this geometric approach.

Since the train is stationary in Bob's frame, its left end, right end, and middle are represented in Alice's diagram by parallel equilocs of Bob. Since Alice agrees with Bob about what point on the train constitutes its middle, these three parallel lines are equally spaced in Alice's diagram. Two photons, created together in the middle of the train, travel in opposite directions at the same speed. Since the train is stationary in Bob's frame, and both photons also have the same speed in his frame (namely 1 foot per nanosecond—here is where the invariance of the velocity of light enters our story) according to Bob the photons arrive at the two ends of the train at the same time. So if we draw a pair of 45 degree photon lines that start at a point on the trajectory of the middle of the train, representing the trajectories of photons moving toward the front and rear, then the points of intersection of the two photons with the two ends of the train represent simultaneous events in Bob's frame and therefore lie on one of his equitemps. All this is illustrated in part 1 of figure 10.7.

It is then easy to deduce that Bob's equitemps in Alice's diagram must make the same angle with the photon trajectories as his equilocs do, just as Alice's equitemps and equilocs do. One sees this most simply by letting each photon be reflected from its end of the train back to the middle. The resulting collection of photon trajectories (shown in part 2 of figure 10.7) form the four sides of a rectangle. It is evident that all the angles with the same label are equal. Therefore the two photon trajectories passing through the black dot on the left do indeed bisect the angles between Bob's equilocs and equitemps passing through that dot.

This conclusion is identical to the rule for how Alice's equilocs and equitemps are oriented with respect to the photon trajectories. Furthermore, because the common speed of both photons in Bob's frame continues to be 1 foot per nanosecond, two of Bob's equilocs associated with places 1 foot apart in his frame must be the same distance apart in Alice's diagram as two of his equitemps associated with times 1 nanosecond apart in his frame. The rules we set up for the orientation of Alice's equitemps and equilocs, and the relation between their scales, have imposed rules on the equilocs and equitemps that Bob must use, if he wishes to represent events using the same points that Alice uses in her diagram. Importantly, those rules for Bob turn out to have exactly the same form as the rules used by Alice. It is therefore impossible for anybody else to tell which of them made the diagram first, and which of them subsequently imposed his or her own equitemps and equilocs on the other's diagram. This wonderful symmetry is required by the principle of relativity. Seeing it emerge in this way affords a vivid demonstration that the principle of relativity is indeed consistent with the frame independence of the velocity of light.

The fact that Bob's and Alice's equitemps and equilocs are both symmetrically situated about the 45 degree photon lines has as an immediate

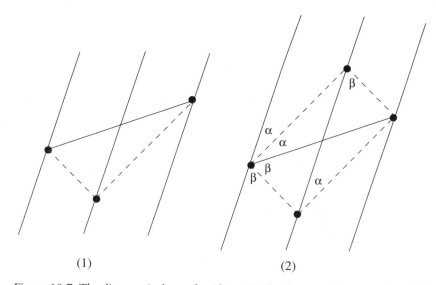

Figure 10.7. The diagram is drawn by Alice. (1) The three equally spaced parallel lines are the two ends and the middle of a train that is stationary in Bob's frame of reference. They establish the direction Bob's equilocs must have in Alice's diagram. The lowest black dot represents the production of two oppositely directed photons at the middle of the train. The dashed lines are the space-time trajectories of the photons. The other two black dots represent the arrival of each photon at an end of the train. Since both photons move at the same speed in Bob's frame of reference and since the train is stationary in Bob's frame, the photons arrive at the ends of the train at the same time in Bob's frame—i.e. the straight line joining the upper two dots is an equitemp for Bob. (2) If the photons are reflected back toward the center of the train when they reach the two ends, they will arrive there at the same time in the event represented by the highest black dot, all four photon lines forming a rectangle. It is evident from the symmetry of the rectangle that the two angles inside the rectangle labled α are equal, as are the two angles inside the rectangle labeled β. Since the two labeled angles outside the rectangle are just spatial translations of two correspondingly labeled angles within it, it follows that both of the photon trajectories passing through the leftmost black dot bisect the angles between Bob's equitemps and equilocs passing through that dot.

consequence the $T = Dv/c^2$ rule for simultaneous events, in the form $T = Dv$ that the rule assumes when one measures times in nanoseconds and distances in feet. This is spelled out in figure 10.8 and its caption.

So Alice and Bob (and Carol and Dick and Eve ...) can all represent events in space and time by the same set of points in a single diagram, on which they each superimpose different families of equilocs and equitemps. The equilocs and equitemps used in any one frame of reference are

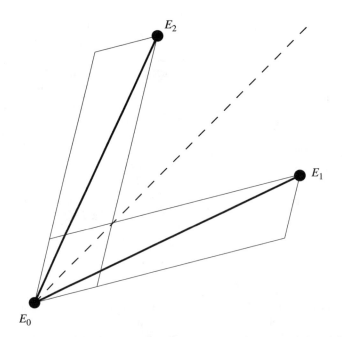

Figure 10.8. The heavy line on which events E_1 and E_0 lie is an equitemp for Bob, and the heavy line on which events E_2 and E_0 lie is an equiloc. The two long, thin parallelograms that overlap in the lower left are made up of segments of equitemps and equilocs for Alice. The entire figure is symmetric when mirrored in the dashed photon line. If Bob moves with speed v in Alice's frame, then since E_2 and E_0 are in the same place according to Bob, according to Alice their separation d in feet is v times their separation t in nanoseconds: $d = vt$. The ratio of d to t is just the ratio of the lengths of the short and long sides of the parallelogram with E_2 and E_0 at opposite vertices. Because of the mirror symmetry in the dashed photon line, v is also the ratio of the short and long sides of the parallelogram with E_1 and E_0 at opposite vertices. But the ratio of the sides of this parallelogram gives the ratio of Alice's time T in nanoseconds separating Bob's simultaneous events E_1 and E_0 to Alice's distance D in feet separating E_1 and E_0. So $T = vD$.

symmetrically disposed about the two fixed perpendicular directions that are the directions along which all photon trajectories are oriented. To put it another way, in any one frame the bisectors of the angles between an intersecting equitemp and equiloc are photon trajectories.

There remains the question of how people using different frames of reference relate their scale factors λ, which give the distance on the page between their equilocs associated with events 1 foot apart and between their equitemps associated with events 1 nanosecond apart. One can acquire substantial insight from appropriately drawn space-time diagrams without ever needing to use the quantitative relation between scale factors, so for now I simply state what the rule is:

The scale factors used by different frames of reference are related by the rule that *unit rhombi used by different observers all have the same area*. Since the altitude of a unit rhombus is the scale factor λ and its base is the scale factor μ (figure 10.9), the analytical expression of this geometric rule is that Alice and Bob's scale factors are related by

$$\lambda_A \mu_A = \lambda_B \mu_B. \tag{10.1}$$

This rule follows simply and directly from the requirement that when Alice and Bob move away from each other at constant velocity, they must each *see* the other's clock running at the same rate, as measured by their own clock. We shall not make use of the rule until later in this chapter, so I defer its elementary derivation to pp. 132–34, after we have examined several examples of uses of space-time diagrams that do not require the explicit form of the connection between Alice's and Bob's scale factors.

Figure 10.10 shows diagrammatically how it is possible for each of two sticks in relative motion to be longer in its proper frame than the other. The two vertical lines represent the space-time trajectories of the left and right ends of a stick. Equilocs in the proper frame of the stick are vertical (since each end of the stick does not change its position in that frame), so equitemps in the proper frame of the stick must be horizontal. Any horizontal slice of the figure shows what things are like at that given moment of time in the frame of the stick.

The two parallel vertical lines that slant upward to the right represent the space-time trajectories of the left and right ends of a second stick. They are equilocs in the proper frame of the second stick. Equitemps in the proper frame of the second stick tilt away from the horizontal by as much as the equilocs tilt away from the vertical. Any slice of the figure with such a tilted equitemp shows what things are like at a particular moment of time in the frame of the second stick.

The horizontal line in figure 10.10 is a particular equitemp in the frame of the first stick. As you look along that line from left to right, you encounter first the left end of the first stick, then the left end of the second stick, then the right end of the second stick, and finally the right end of the first stick. Thus in the proper frame of the first stick the two ends of the first stick extend beyond the two ends of the second stick: the second stick is shorter than the first.

On the other hand, the tilted line is a particular equitemp in the frame of the second stick. As you move along that tilted line from lower left to upper right, you encounter first the left end of the second stick, then the left end of the first stick, then the right end of the first stick, and finally

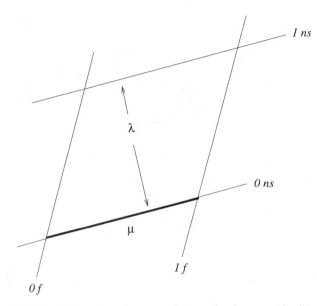

Figure 10.9. The unit rhombus for some frame of reference. The lines labeled 0 ns and 1 ns represent events 1 nanosecond apart and the lines labeled 0 f and 1 f represent events 1 foot apart. Because the distance in the diagram between the two equitemps, regarded as the height of the rhombus, is the scale factor λ, and because the heavier portion of the lower equiloc, regarded as the base of the rhombus, is the scale factor μ, the area of the rhombus—its base times its height—is just the product $\lambda\mu$.

the right end of the second stick. Thus in the proper frame of the second stick the two ends of the second stick extend beyond the two ends of the first stick: the first stick is shorter.

What the figure makes explicit is that if two sticks are in motion relative to one another, then their comparative lengths depend on the convention one employs for the simultaneity of events in different places. The various pieces of a stick (its two ends, its middle, a point two-thirds of the way along the stick, etc.) are situated in different places. Which parts of the space-time trajectories of each piece of the stick one puts together to make up what one would like, at a given moment of time, to call *the stick* depends on which events in the history of each of those spatially separated pieces of stick one chooses to regard as simultaneous.

What is independent of any such convention is the totality of all the space-time trajectories of all the pieces of both sticks. What is conventional and frame-dependent is how one chooses to slice those trajectories with equitemps to form the *stick-at-a-given-moment*.

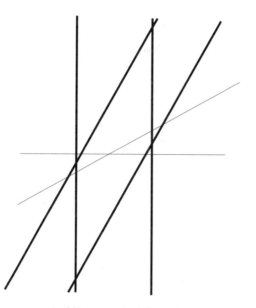

Figure 10.10. The two vertical lines are the left and right ends of a stick. The two parallel lines that tilt upward to the right are the left and right ends of a second stick that moves to the right past the first stick. The horizontal line is an equitemp in the frame in which the first stick is stationary. The line that tilts upward to the right is an equitemp in the frame in which the second stick is stationary. It tilts away from the horizontal by the same amount that the lines representing the ends of the second stick tilt away from the vertical.

Notice that there is a third frame of reference (moving to the right with respect to the first stick at a speed less than the second) in which both sticks have the same length. That frame is the one with equitemps parallel to the line that joins the point of intersection of the trajectories of the left ends of the sticks to the point of intersection of the trajectories of the right ends.

Similarly, figure 10.11 shows diagrammatically how it is possible for each of two clocks, in relative motion, to run faster than the other in its proper frame. The vertical row of numbered circles represents seven moments in the history of a clock and the reading of the clock at those moments. The slanting row represents six moments in the history of a second clock, moving to the right relative to the first and its reading at those moments. Both clocks are in the same place at the same time when they read 0 and are therefore represented at that moment in their histories by one and the same circle. Everybody, regardless of what frame of ref-

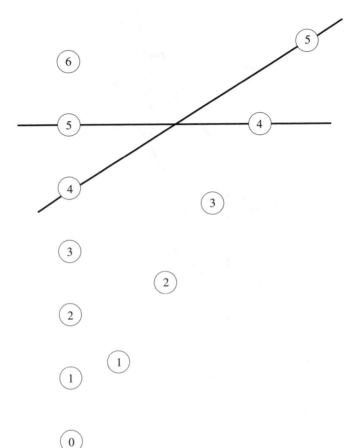

Figure 10.11. Several moments in the histories of two uniformly moving clocks. Each point that represents a clock has been expanded to a circle that displays the reading of that clock. Both clocks read 0 at the same place and time. That event is represented by just a single circle. Subsequent readings of the first clock (1–6) are shown on the set of circles uniformly spaced along a vertical line; subsequent readings of the second clock (1–5) are shown on the set of circles that lie on a line sloping upward to the right. The horizontal line is an equitemp in the frame of the first clock. The slanting line is an equitemp in the frame of the second clock, so it tilts away from the horizontal by the same amount that the line of pictures of the second clock tilts away from the vertical.

erence they use, agrees that both clocks read 0 at the same time, because the clocks read 0 at the same time *and* in the same place.

Equilocs in the proper frame of the first clock are vertical (since the line on which the seven moments in the history of the first clock lie is vertical), so equitemps in the proper frame of the first clock are horizontal. Since the events in which the second clock reads 4 and the first clock reads 5 lie on a single horizontal line, those two readings happen at the same time, in the proper frame of the first clock. Since both clocks also read 0 at the same time, the second clock is running at $\frac{4}{5}$ the rate of the first, according to the proper frame of the first clock.

Equilocs in the proper frame of the second clock are parallel to the line on which the six moments in the history of the second clock lie, so equitemps in the proper frame of the second clock make the same angle with the horizontal as that line of clocks makes with the vertical. Such an equitemp is shown connecting the events in which the first clock reads 4 and the second clock reads 5. Since both clocks also read 0 at the same time, the first clock is running at $\frac{4}{5}$ the rate of the second, according to the proper frame of the second clock.

The figure makes explicit the fact that a comparison of the rates of two clocks in relative motion depends crucially on the convention one adopts for the simultaneity of events in different places. Because each frame uses a different convention for how to slice space-time up into equitemps, there is no contradiction in each frame maintaining that clocks in the other frame run slowly compared with its own clocks.

If we stopped with figure 10.11, the question of which clock was *actually* running slower would be a matter of convention, empty of real content, depending on whose notion of simultaneity you chose to adopt. Suppose, however, that the second clock suddenly reverses its direction of motion and returns to the first. One can then compare the clocks directly when they are back at the same place at the same time and see which has advanced by the greater amount.

In thinking about this it is essential to recognize that the process of turning around breaks the symmetry between the two clocks. The first clock is stationary in a single inertial frame of reference throughout its entire history. The proper frame of the second clock, however, changes from one inertial frame of reference, moving uniformly to the right, to another, moving uniformly to the left, at the moment it turns around. There is no single inertial frame of reference in which the second clock is stationary throughout its history, and the enormous decelerations and accelerations attended upon turning around and heading back to the first clock will be quite evident to anybody moving with the second clock.

In the frame of reference of the first clock (which uses horizontal equitemps), it is clear from figure 10.12 that when the trip is over the second clock, running slowly for the entire journey, will have advanced only by 8 (4 on the outward journey and 4 on the inward journey) while the first clock has advanced by 10. So when the two clocks come back together, the first will read 10 and the second, 8, as indicated in the figure.

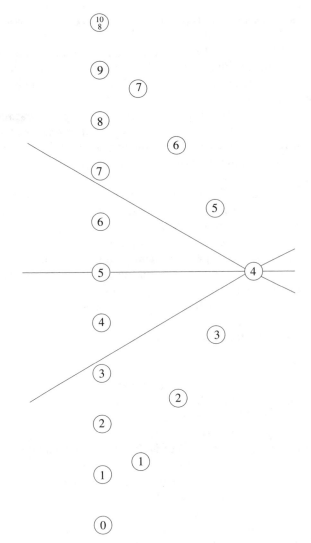

Figure 10.12. Two identical clocks. The first is shown at eleven different moments along a vertical line, as its reading advances from 0 to 10. The second moves uniformly away from the first as its reading advances from 0 to 4; it then moves uniformly back to the first, as its reading advances from 4 to 8. At the bottom and top of the figure, both clocks are at the same place at the same time and are represented by a single circle. The first clock is stationary in a single inertial frame of reference. Since equilocs are vertical in that frame, equitemps are horizontal. The horizontal line is an equitemp in the frame of the first clock. The line slanting upward to the right is an equitemp in the frame of the second clock as it moves away from the first. The line slanting upward to the left is an equitemp in the (different) frame of the second clock as it moves back to the first.

Things are trickier from the point of view of the second clock, since two different inertial frames of reference are involved. In the frame moving outward with the second clock, the first clock runs slowly and advances only by 3.2 (from reading 0 to reading 3.2) during the time the second clock advances by 4 (from 0 to 4). This is revealed by the lower of the two tilted equitemps in figure 10.12. Similarly, in the frame moving inward with the second clock, the first clock is also running slowly and advances only by 3.2 (from reading 6.8 to reading 10) as the second clock advances by 4 (from 4 to 8), as revealed by the upper tilted equitemp in figure 10.12.

The indisputable fact that the first clock reads 10 and the second reads only 8 when they are reunited makes sense from the point of view of the second clock, even though the first clock runs slowly in both the outgoing and the incoming frames. The missing 3.6 units of first-clock time ($3.6 = 10 - 2 \times 3.2$) comes from a correction that must be made in the notion of *what-the-first-clock-reads-now* when the second clock changes frames. As figure 10.12 shows, at the place and time of turnaround, when the second clock reads 4, the faraway first clock *now* reads 3.2 according to the notion of simultaneity in the outgoing frame, but it *now* reads 6.8 according to the notion of simultaneity in the incoming frame. It is this adjustment, with the change of frames, of what the first clock is doing *now*, that accounts for the missing time. The adjustment is reminiscent of, but considerably more subtle than, the way one has to reset one's watch to the new time zone, when one emerges from an airplane that has flown across the Atlantic.

The essential but uncomfortably artificial role played in figure 10.12 by the different simultaneity conventions in different frames of reference drops out of the story if we ask not what people moving with each clock *say* about the current reading of the other clock, but what they actualy *see* it doing. Figure 10.13 reproduces the clocks of figure 10.12, without the equitemps appropriate to the three different frames of reference, but with the trajectories (dashed lines) of photons emitted by each clock as its reading changes. Since the slowing-down factor for the moving clocks is $\frac{4}{5}$, the relative velocity of the clocks is $v = \frac{3}{5}c$, and therefore the Doppler factor discussed in chapter 7, $\sqrt{\frac{1+v/c}{1-v/c}}$, is 2. People watching a clock moving away from them (or moving away from the clock) at $\frac{3}{5}$ the speed of light will *see* it running at half its proper rate; people watching a clock moving toward them (or moving toward the clock) at $\frac{3}{5}$ the speed of light will *see* it running at twice its proper rate. Note that the factors of 2 and $\frac{1}{2}$ extracted from the formula for the Doppler factor emerge automatically from the geometry of figure 10.13.

When the light emitted by the second clock, as it changes to 1, 2, 3, and 4, reaches people with the first clock, the first clock reads 2, 4, 6,

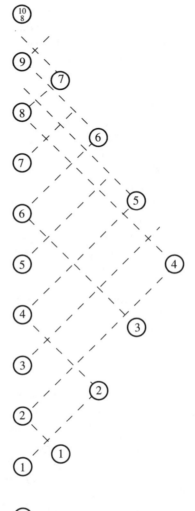

Figure 10.13. This shows again the two clocks of figure 10.12, but without the equitemps, and with many photon trajectories indicating what somebody moving with either clock *sees* the other clock to be doing. Each clock emits a flash of light each time it shows a new integer, and those flashes are seen by people moving with the other clock.

and 8. When light emitted by the second clock as it changes to 5, 6, 7, and 8 reaches people with the first clock, the first clock reads 8.5, 9, 9.5, and 10. So the people with the first clock see the second clock running at half its proper rate for 80 percent of their time and at twice its proper rate for 20 percent of their time. The much greater time seen running slowly overwhelms the rather brief time seen running fast, and the net effect is that the second clock has not advanced as much as the first when the journey is over. On the other hand when the light emitted by the first clock, as it changes to 1 and 2, reaches people moving with the second clock, the second clock reads 2 and 4. When light emitted by the first clock as it changes to 3, 4, 5, 6, 7, 8, 9, and 10 reaches people moving with the second clock, the second clock reads 4.5, 5, 5.5, 6, 6.5, 7, 7.5, and 8. So people with the second clock see the first clock running at half its proper rate for half the time and at twice its proper rate for half the time. Since it is seen running at twice its proper rate for half the time, this already insures that it will have advanced by as much as the second clock when they are back together, and because it is seen running at half its proper rate for the remaining half, it must have advanced by an additional 25 percent when they are back together, as indeed it has.

The fact that two identical clocks, initially in the same place and reading the same, can end up with different readings, if they move apart from each other and then back together, is sometimes called the *clock paradox* or, even more commonly, the *twin paradox*. If the motion of one clock during their separation were the mirror image of the motion of the other, this would indeed be a paradox, since there would be no way to determine which clock advanced the most during the journey. But there is no paradox if the two clocks move asymmetrically during their separation, as they do in the example we have just examined, in which one clock moves uniformly at all times while the other suffers an abrupt change in velocity when its proper frame changes from the outgoing inertial frame to the incoming frame.

The "twin paradox" refers to the dramatic version of the phenomenon, arising when the two "clocks" are identical twins. If one twin goes to a star 3 light years away in a super rocket that travels at $\frac{3}{5}$ the speed of light, the journeys out and back each take 5 years in the frame of the earth. But since the slowing-down factor is $\sqrt{1 - \left(\frac{3}{5}\right)^2} = \frac{4}{5}$, the twin on the rocket will age only 4 years on the outward journey, and another 4 years on the return journey. When she gets back home, she will be 2 years younger than her stay-at-home sister, who has aged the full 10 years. This is unfamiliar, since nobody has ever gone anywhere at so phenomenal a speed, but not paradoxical. We shall examine an entertaining addendum to the clock paradox in our brief glimpse of general relativity in chapter 12.

As a variation on the apparent paradox of mutual length contraction, consider a situation in which Alice runs toward the front door of a long narrow barn that stretches away from her. She carries a long horizontal pole that points toward the door. The proper length of the pole is greater than the proper length of the barn, so if the pole were stationary it could not fit in the barn. But Alice runs so fast that in the barn frame the shrinking of the pole makes it shorter than the barn, and the pole fits comfortably into the barn. In Alice's frame, on the other hand, the pole retains its proper length, and it is the barn, moving toward Alice, that suffers a length contraction, so it is even more impossible (if something impossible can become more impossible) for the pole to fit in the barn.

What's going on here? To avoid having to worry about the pole crashing into the wall of the barn opposite the entrance, or the considerable complications introduced by the difficult process of quickly slowing down and stopping a pole that moves at a speed comparable to the speed of light, let's suppose that the barn has a rear door as well as a front door, so the pole can continue moving uniformly out of the barn, without ever changing its speed. Is there or is there not a time when the pole is in the barn?

Notice the appearance in this question of the crucial word "time." The resolution of this apparent paradox is that whether or not a moving pole fits in a barn does indeed depend on the frame of reference. For "the pole is in the barn" really means "all the parts of the pole are between the front and rear door of the barn at the same time." Since different parts along the pole are in different places, and since different frames of reference use different conventions in determining whether events in different places are simultaneous, there can indeed be a legitimate disagreement about whether there are moments of time at which all the different parts of the pole are between the two doors of the barn.

The space-time diagram in figure 10.14 makes clear what is really going on. Suppose that each barn door is closed except when the pole is actually in the doorway. The two vertical lines represent these doors, which are shut when the lines are solid and open when they are dotted. The points making up the moving pole lie in the gray shaded region, which is bounded on the left by the space-time trajectory of the left end of the pole and on the right by the space-time trajectory of the right end.

Equitemps in the barn frame are horizontal. Two are shown. The lower horizontal equitemp contains the moment when the rear end of the pole has entered the barn and demonstrates that at that moment the front end of the pole has yet to leave the barn. The upper horizontal line contains the moment when the front end of the pole is about to leave the barn and demonstrates that at that moment the rear end of the pole is well within the barn. At all barn-frame times between those associated with the two

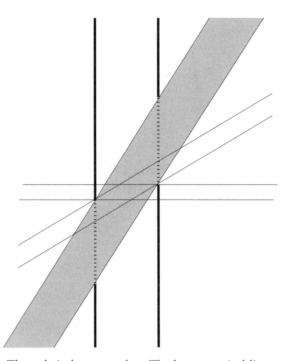

Figure 10.14. The pole-in-barn paradox. The heavy vertical lines are the space-time trajectories of the left and right doors of the barn, which are shut when the lines are solid and open when the lines are dotted. The lines bounding the shaded gray region are the space-time trajectories of the left and right ends of the pole. Interior points of the gray region represent the interior points of the pole. The two horizontal lines are equitemps in the barn frame. They demonstrate that there is a range of barn-frame times when the pole is entirely inside the barn and both doors are shut. The two lines slanting up to the right (but less steeply than the lines representing the ends of the pole) are equitemps in the pole frame. They demonstrate that there is a range of pole-frame times when the pole extends all the way through the barn and through both (open) barn doors.

horizontal lines, the pole is entirely inside the barn, and both barn doors are shut.

But equitemps in the pole frame tilt upward to the right. Two such lines are shown. The lower tilted equitemp contains the moment when the front end of the pole is about to leave the barn and demonstrates that at that moment the rear end of the pole has not yet entered the barn. The upper tilted equitemp contains the moment when the rear end of the pole finally enters the barn. At that moment of pole-frame time the front end has already left the barn. There is no moment of pole-frame time

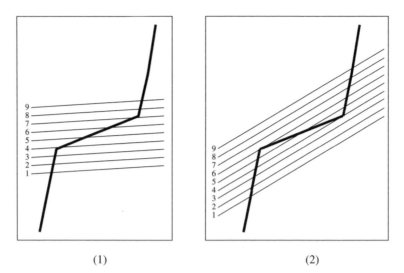

(1)   (2)

Figure 10.15. The heavy line that bends twice is the trajectory of an object. The middle part of the line is more horizontal than vertical, indicating that for that part of its trajectory the object is moving faster than light. Part 1 shows several equitemps imposed on the diagram by Alice; part 2 shows several equitemps imposed on the same diagram by Bob.

during which both barn doors are shut. In the pole frame, the pole is never entirely within the barn.

The important lesson of the pole-in-barn paradox is that even so innocent sounding a sentence as "The pole is shut up in the barn" can involve an implicit judgment about the simultaneity of events in different places when it is applied to a moving pole.

Space-time diagrams also make it quite easy to examine the peculiarities of motion faster than light. Figure 10.15 shows the trajectory (heavy line) of an object that initially moves slower than light, then suffers a period of faster-than-light motion, and then returns to its initial velocity. You can see that the middle segment of the line does indeed tilt away from the vertical at an angle even greater than the 45 degree angle of photon lines, while the two outer segments tilt away from the vertical at less than 45 degrees.

Part 1 imposes on the trajectory the equitemps of Alice, which confirm the above description. The lines are numbered in order of increasing time. At times 1–3, the object moves slower than light; at times 4–7, it moves faster than light; and at times 8–9, it is back to moving at its original velocity.

Part 2 imposes on the trajectory the equitemps of Bob, who is moving to the right faster than Alice, though not, of course, faster than light (as you can confirm by noting that his equitemps tilt up from the horizontal at less than 45 degrees). Bob tells a more interesting story than Alice. At times 1–3, an object on the left of the figure moves at a speed less than light. At time 4, two more objects appear out of nothing on the right of the figure! One of them moves toward the left side of the figure at a speed greater than light, while the other moves further to the right at a speed less than light. Between times 4 and 6, all three objects are present, but at time 6 the faster-than-light object meets the slower-than-light object on the left and both disappear, leaving only the slower-than-light object on the right to continue peacefully on at times 7–9.

Strange as it may be, there is nothing wrong in principle, with a disagreement about how many objects are present at a given time, since "at a given time" means different things in different frames, as the diagrams make quite explicit. Rather more disturbing is the diagreement between Alice and Bob over the order of events in the history of the faster-than-light object. Alice says it originates on the left when the slower-than-light object suddenly acquires an enormous speed and terminates on the right when the object abruptly drops back to a slower-than-light speed. Bob disagrees, saying that the faster-than-light object is born with the slower-than-light object on the right before it disappears together with the slower-than-light object on the left.

We have encountered this problem in a particular case in chapter 9 (figure 9.9). The diagrammatic analysis shows that it is always a problem: if an object moves faster than light then there will necessarily be frames of reference that disagree about the temporal order of events in its history, depending on whether their equitemps tilt upward less steeply or more steeply than the space-time trajectory of the object. (It is only for faster-than-light trajectories that it is possible for a frame of reference to have the second kind of equitemp.) As we noted in chapter 9, the difficulties with such a disagreement are overwhelming if the object is capable of revealing the direction of time by aging, rotting, burning up, etc.

Figure 10.16 reveals yet another problem with objects moving faster than light. The light solid line on the left side of the figure is the space-time trajectory of Alice. The black circles labeled 1 and 2 are two events in her history. Event 1 happens before event 2. When event 2 takes place Alice sends a faster-than-light signal to Bob (white circle on the left), who is moving away from her at a prodigious speed, close to but slower than the speed of light, so it is possible for her faster-than-light signal (the higher heavy line) to catch up with him. The light solid line slanting downward to the left from Bob at the moment Alice's signal reaches him is an equitemp for Bob. Note that it is (slightly) more horizontal than

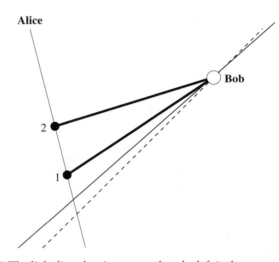

Figure 10.16. The light line slanting upward to the left is the space-time trajectory of Alice. The two black circles are events at which she is present. When event 2 takes place Alice sends a faster-than-light message (upper heavy line) to Bob (somewhat larger white circle). When Bob receives Alice's message he sends the information back to Alice, using a faster-than-light signal of his own (lower heavy line). The signal reaches Alice at the moment event 1 takes place, even though event 1 takes place before event 2 (in everybody's frame). The dashed line passing through the white circle is a photon line, which demonstrates that the lighter line passing through the white circle can indeed be an equitemp in Bob's frame, if Bob is moving fast enough to the right.

the dashed photon line emanating from Bob at the same moment, so it is indeed a valid equitemp. The figure demonstrates that when Bob receives Alice's signal from event 2, event 1 has not yet happened in his frame, since it lies above Bob's equitemp. (But notice that Bob agrees with Alice that event 1 happens before event 2; there can be disagreement about temporal order only for space-like separated events.) So if Bob can also send faster-than-light signals, then he can send one (the lower heavy line) back to Alice that arrives at the moment of event 1.

In this way Alice can send herself messages into the past. Suppose event 2 is the end of a race and event 1 is the last opportunity to place a bet on the race. When the race ends Alice sends Bob the name of the winner using a faster-than-light signal. When Bob gets the name he sends it back to Alice using his own faster-than-light signal. Alice receives it before the race begins, so she can place a certain bet on the winner. (In Alice's frame, of course, the name of the winner magically appears out of nowhere and races off to Bob at faster-than-light speed, but she knows that the reason this has happened is that after the race is over she will send it to Bob with her own faster-than-light signal.)

A strange business, of course. Sending messages into the past to induce the recipient to change the course of events that have already taken place may be standard fare in science fiction tales, but it is hard to reconcile with a coherent picture of the real world. The possibility it would open up of signaling into the past is one of the stronger arguments against faster-than-light motion. To save the possibility of such motion while avoiding signaling into the past, one must add yet another proviso: anything capable of moving faster than light cannot be capable of being produced at will. It must simply come into existence, or flash past, in a random uncontrollable way, which makes it unusable for signaling.

The relations of space-like, time-like, or light-like separation between two events, developed in chapter 8, are elementary features of the representations of the events in a diagram. Figure 10.17 shows an event, E, and the trajectories of two oppositely moving photons, present at E. Points in the two regions labeled "space-like" represent events that are space-like separated from E, since a line joining any point of these regions to E is more horizontal than vertical. It can therefore be an equitemp in some frame of reference. In that frame the event happens at the same time as E. Points in the regions labeled "time-like" give events that are time-like separated from E, since a line joining such a point to E is more vertical than horizontal and can therefore be an equiloc in some frame of reference. In that frame the event happens in the same place as E. The photon trajectories forming the boundaries of these regions contain the events that are light-like separated from E.

The two regions time-like separated from E can be further divided into the future and past of E, since all frames of reference agree on the time order between time-like separated events. But an event that is space-like separated from E cannot be unambiguously assigned to either its future or its past, since different frames will disagree on whether the event happened before or after E, as illustrated in figure 10.18.

Figure 10.19 shows perhaps the simplest way of seeing why the interval between two events does not depend on the frame of reference in which the time T and distance D between the events are evaluated.

The figure is drawn by Carol, who uses a frame of reference in which Bob and Alice move with the same speed but in opposite directions, Bob moving to the left and Alice, to the right. There are three important consequences of their speeds being the same in Carol's frame. (a) Their scale-factors λ (and their scale factors μ) must be the same (by symmetry) in Carol's diagram. (b) Bob's equilocs must be rotated counterclockwise from the vertical by the same angle as Alice's equitemps are rotated counterclockwise from the horizontal, so they are perpendicular. (c) In the same way, Bob's equitemps are perpendicular to Alice's equilocs.

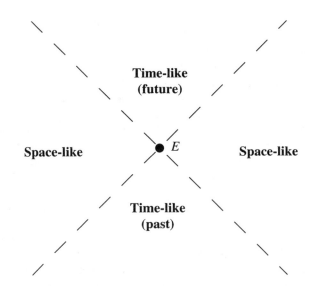

Figure 10.17. The black circle is an event, E. The two dashed lines are the trajectories of photons present at E and moving in opposite directions. The photon trajectories divide the diagram into four regions. Two contain the events that are space-like separated from E, and the other two contain the time-like separated events.

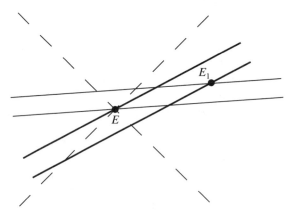

Figure 10.18. The event E from figure 10.17 and a second event E_1, space-like separated from E. The two heavier solid parallel lines are equitemps for a frame in which E_1 occurs before E, since the equitemp containing E_1 is lower in the diagram than the one containing E. But the pair of lighter solid lines are equitemps for a frame in which E_1 occurs after E, since the equitemp containing E_1 is higher in the diagram than the one containing E.

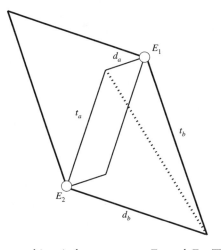

Figure 10.19. The two white circles are events E_1 and E_2. The large parallelogram having the two circles on opposite vertices is made up of segments of Bob's equitemps and equilocs, whose lengths are t_b and d_b. The smaller parallelogram with the circles on opposite vertices consists of segments of Alice's equitemps and equilocs, of lengths t_a and t_b. The diagram is drawn by Carol, in whose frame Bob and Alice have equal speeds in opposite directions. Note the two right triangles, both having the same (dotted) line as hypotenuse, that demonstrate that $t_a^2 + d_b^2 = t_b^2 + d_a^2$.

Keeping this in mind, consider the white circles in figure 10.19, which represent events E_1 and E_2. The large parallelogram having E_1 and E_2 at opposite vertices is made up of Bob's equitemps and equilocs, having lengths t_b and d_b; the small parallelogram consists of Alice's equitemps and equilocs, having lengths t_a and d_a. Because of the perpendicularity just noted, there are two right triangles in the figure with a common hypotenuse (the dotted line). The lower triangle has sides of length t_a and d_b. The upper one has sides of length t_b and d_a. Since they have the same hypotenuse, Pythagoras tells us that $t_a^2 + d_b^2 = t_b^2 + d_a^2$, or

$$t_a^2 - d_a^2 = t_b^2 - d_b^2. \tag{10.2}$$

Now distances in the diagram are related to the times and distances between the events by $t_a = \mu_a T_a$, $d_a = \mu_a D_a$, $t_b = \mu_b T_b$, and $d_b = \mu_b D_b$. But since $\mu_a = \mu_b$, it follows from (10.2) that

$$T_a^2 - D_a^2 = T_b^2 - D_b^2, \tag{10.3}$$

which is the statement that the interval between the events is independent of whether one calculates it using Alice's or Bob's times and distances.

> We conclude with an experiment that establishes how the scale factors λ_A or μ_A used by Alice are related to those, λ_B or μ_B, used by Bob. In the diagrammatic representation of the experiment, this relation has a simple geometric expression. With this in hand, we can give an equally simple geometric interpretation of the invariant interval, and use it to show geometrically how to measure the interval between two events with a single clock, as described, but not proved, at the end of chapter 8.

The connection between the scales Alice and Bob use on their equilocs (or equitemps) follows directly from the diagrammatic representation of what they see when each watches the other's clock. We now show that an equiloc of Alice separating events a time T apart in her frame is related to an equiloc of Bob separating events the same time T apart in his, by the following rule: *The two rectangles of photon trajectories having the two lines for diagonals have the same area.*

The rule is illustrated in figure 10.20. Part 1 shows two moments at which a clock, stationary in Alice's frame, reads 0 and T. The two moments in the history of the clock lie on an equiloc of Alice a distance $\mu_A T$ apart. The two photon trajectories emerging from the lower picture of the clock and the two entering the upper picture of the clock form a rectangle—a *light rectangle*—which has as its diagonal the segment of Alice's equiloc connecting the clocks. Part 2 of figure 10.20 shows the same construction for a clock stationary in Bob's frame. The length $\mu_B T$ of the segment of Bob's equiloc connecting the two moments in the history of his clock exceeds the length $\mu_A T$ of the corresponding segment associated with Alice's equiloc. But the areas of the two surrounding light rectangles are exactly the same.

To see why, take the case in which the two clocks are in the same place when they read 0. This is illustrated in figure 10.21, which results from sliding (without rotating) part 2 of figure 10.20 over to part 1, to bring the two clocks reading 0 into coincidence, and then adding two more events and some labels. Let Alice, moving with her clock, look at Bob's clock at the moment hers reads T, and let Bob, moving with his clock, look at Alice's at the moment his own reads T. Each will see the other's clock reading the same earlier time t, because the relations among Alice and Bob and their clocks are completely symmetrical: each looks at the other's clock after the same time T has passed on their own clock; each regards the other's clock as moving away at the same speed; and for each the speed of the light coming from the other's clock is 1 foot per nanosecond.

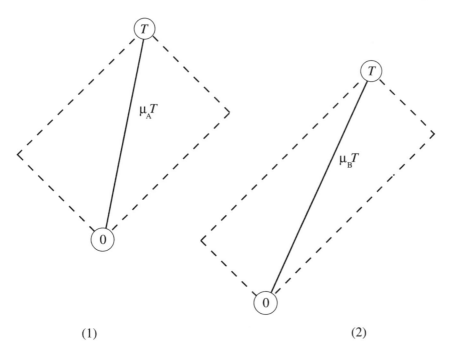

(1)   (2)

Figure 10.20. (1) An equiloc in Alice's frame of reference, separating two events that are a time T apart in Alice's frame. The line can be viewed as the space-time trajectory of a clock that is stationary in Alice's frame, reading 0 at the first event and T at the second. (2) The same as part 1, but for a different pair of events and different clock that is stationary in Bob's frame of reference. Note that the line connecting the events in which the clock stationary in Bob's frame reads 0 and T, which has length $\mu_B T$, is longer than the corresponding line in Alice's frame, which has length $\mu_A T$—i.e. Alice and Bob use different scale factors μ to relate separation in time to distance along their equilocs. The ratio of their scale factors is just the ratio of these two lengths. Although the scale factors differ, the *areas* of the two light rectangles formed by photon trajectories emerging from the events are the same. This is established in figure 10.21.

A glance at figure 10.21 reveals that the ratio b/a of the short side of Bob's light rectangle to the short side of Alice's is the same as the ratio $\mu_A t/\mu_A T = t/T$. And the ratio A/B of the long side of Alice's light rectangle to the long side of Bob's is the same as the ratio $\mu_B t/\mu_B T = t/T$. So

$$A/B = t/T = b/a, \tag{10.4}$$

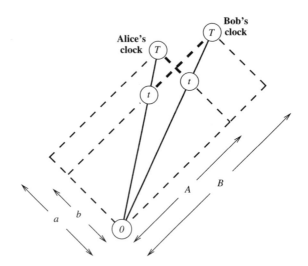

Figure 10.21. The two parts of figure 10.20 have been slid together (without rotating either part) to describe the situation in which the two clocks read 0 at the same place and time. At the moment each clock reads T, somebody with that clock looks at the other clock and sees it reading t, as indicated by the two heavier segments of the photon lines.

and therefore

$$Bb = Aa. \tag{10.5}$$

The left side of (10.5) is the area of Bob's light rectangle and the right side is the area of Alice's. This is what we wished to establish.

This equality of the areas of the light rectangles implies the equality (10.1) of the product $\lambda\mu$ of scale factors. For four copies of either of the two identical triangles making up either rectangle in part 1 of figure 10.22 can be reassembled into a rhombus whose sides have length μT and are a distance λT apart, as shown in part 2 of figure 10.22. The area of the rhombus is $\lambda\mu T^2$, so the area of the light rectangle is $\frac{1}{2}\lambda\mu T^2$, where one uses λ_A and μ_A for Alice's rectangle and λ_B and μ_B for Bob's. Since Alice's and Bob's rectangles have the same area, this establishes that the product $\lambda\mu$ is independent of frame of reference. It also shows that the relation between the area A of either rectangle in figure 10.20 and the time T between the clock present at the two events on its opposite corners is just

$$T^2 = A/(\tfrac{1}{2}\lambda\mu). \tag{10.6}$$

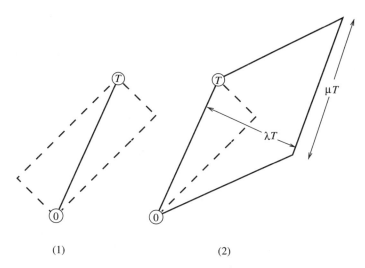

(1) (2)

Figure 10.22. The area of either of the light rectangles in figure 10.20, shown here in part 1, is half the area of the rhombus shown in part 2, since the rhombus can be assembled out of two copies of each of the two right triangles that make up the rectangle. Since the area of the rhombus in part 2 is the length μT of a side times the distance λT between sides, the area of the light rectangle in part 1 is $\frac{1}{2}\lambda\mu T^2$.

Abstracting from this, we can conclude that the area A of *any* rectangle of photon trajectories with two time-like separated events at opposite vertices is just a frame-independent product of scale factors,

$$A_0 = \tfrac{1}{2}\lambda\mu, \tag{10.7}$$

times the square of the time T between the two events in the frame in which they happen at the same place:

$$A = A_0 T^2. \tag{10.8}$$

But T^2, the square of the time between two events in the frame in which they happen at the same place, was defined in chapter 8 to be the *squared interval* I^2 between the events. Thus the squared interval between two time-like separated events is just the area (in units of A_0, the frame independent area of the unit light rectangles) of the light rectangle having the events at opposite vertices:

$$I^2 = A/A_0. \tag{10.9}$$

Because of the explicit symmetry of the diagrams under the interchange of space and time, we can also conclude that the area A of the rectangle of photon trajectories with two space-like separated events at opposite vertices is A_0 times the square of the distance D between the two events in the frame in which they happen at the same time. But D^2 was defined in chapter 8 to be the squared interval between two events when their separation is space-like. Therefore (10.9) also gives a geometric representation of the interval between two space-like separated events, in terms of the area of the light rectangle having the events at opposite vertices.

This geometric interpretation of the interval enables one to see directly from figure 10.23 that the squared interval between two time-like separated events is the difference of the square of the time between the events and the square of the distance between them, regardless of the frame in which that time and distance are evaluated. The solid lines are an equi-temp and equiloc in Alice's frame that connect the events E_1 and E_2 to a third event, E_3. The squared interval I^2 between events E_1 and E_2 is proportional to the area $(a - c)(b + d)$ of the light rectangle with the events at opposite vertices, which, because $ad = bc$ (as demonstrated in the caption of figure 10.23), is just $ab - cd$:

$$I^2 = (ab - cd)/A_0. \qquad (10.10)$$

But (see figure 10.23) ab is proportional to the squared interval between the events E_1 and E_3, while cd is proportional to the squared interval between E_2 and E_3. Since E_1 and E_3 happen in the same place in Alice's frame, the squared interval between them is T^2, the square of Alice's time between them; since E_2 and E_3 happen at the same time in Alice's frame, the squared interval between them is D^2, the square of Alice's distance between them. But since E_3 happens in the same place as E_1 and at the same time as E_2 in Alice's frame, T and D are also Alice's time and distance between E_1 and E_2. Since

$$T^2 = ab/A_0, \quad D^2 = cd/A_0, \qquad (10.11)$$

it follows from (10.10) that

$$I^2 = T^2 - D^2. \qquad (10.12)$$

The analogous conclusion for space-like separated events, $I^2 = D^2 - T^2$, is demonstrated in the same way from the figure given by reflecting figure 10.23 in any of the 45 degree photon lines slanting upward to the right.

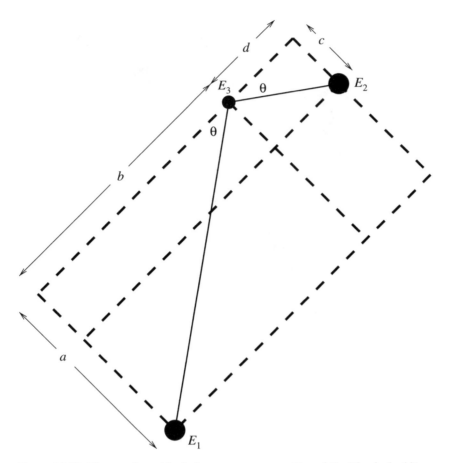

Figure 10.23. The two large black dots are two events E_1 and E_2. The dashed lines are photon lines. The two solid lines are equitemps and equilocs in Alice's frame of reference, which therefore make the same angles with either of the photon lines passing through a third event, E_3. As a result, the right triangle with sides d and c is just a scaled-down version of the right triangle with sides a and b. Because the right triangles are similar, $a/b = c/d$ and therefore $ad = bc$.

The proof of (10.12) based on figure 10.23 requires a tiny amount of algebra. An alternative proof that is entirely geometrical is given in figure 10.24. Part 1 of figure 10.24 reproduces figure 10.23, except that each of the three light rectangles having a line between events as its diagonal has been replaced by a rhombus of twice the area, with the line as one of its four sides. Part 2 reassembles the two smaller rhombi and two copies of the triangle with the three events E_1, E_2, and E_3 at its vertices, into a quadrilateral. Part 3 reassembles the largest rhombus and two copies

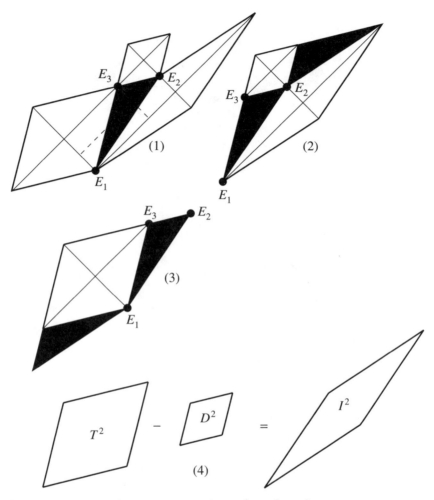

Figure 10.24. A purely geometric proof that $I^2 = T^2 - D^2$.

of the same triangle into exactly the same quadrilateral. Therefore the areas of the three rhombi are related as shown in part 4, which establishes (10.12) as a purely geometric result, since the three areas are proportional to I^2, T^2, and D^2.

We can also use space-time diagrams to derive the efficient way to measure the interval between two events using only light signals and a single clock, described without proof at the end of chapter 8. Figure 10.25 shows the case in which the events E_1 and E_2 (the gray circles) are time-like separated. The heavy solid line is the world line of Alice's clock, which is present at the event E_1 where it reads t_1. The dashed photon line going

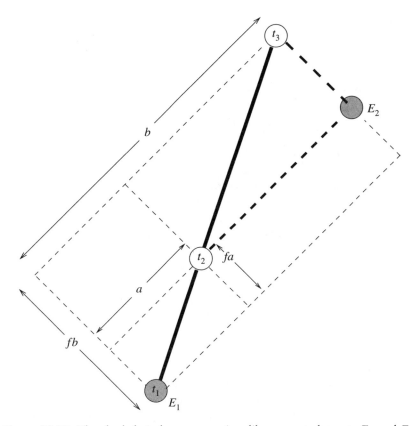

Figure 10.25. The shaded circles are two time-like separated events E_1 and E_2. The heavy solid line is a clock moving uniformly with Alice that is present at E_1 when it reads t_1. The lower heavy dashed photon line demonstrates that somebody present at the event E_2 sees Alice's clock reading t_2 when E_2 takes place. The upper heavy dashed photon line demonstrates that Alice's clock reads t_3 when Alice sees E_2 taking place.

down to the left from E_2 to Alice's clock reveals that Bob, who is present at E_2 and looking at Alice's clock, sees it reading t_2 when E_2 takes place. The dashed photon line going up to the left from E_2 to Alice's clock, reveals that when Alice sees E_2 happening her clock reads t_3.

The squared interval between Alice's clock reading t_3 and t_1 is just $1/A_0$ times the area of the large light rectangle whose sides are labeled b and fb. Since the squared interval between two time-like separated events is just the square of the time between them according to a uniformly moving clock present at both, we have

$$(t_3 - t_1)^2 = fb^2/A_0. \tag{10.13}$$

For the same reason, we have

$$(t_2 - t_1)^2 = fa^2/A_0, \tag{10.14}$$

(where we have used the fact that the light rectangle with t_2 and t_1 on opposite vertices in figure 10.25 is just a scaled-down version of the larger one with t_3 and t_1 at opposite vertices, so that the ratio f of the short to the long side of both rectangles is the same). Finally, the squared interval I^2 between the two events of interest, E_1 and E_2, is $1/A_0$ times the area of the very long light rectangle with sides b and fa:

$$I^2 = fab/A_0. \tag{10.15}$$

Comparing (10.15) with (10.13) and (10.14), we see that

$$I^2 = (t_3 - t_1)(t_2 - t_1), \tag{10.16}$$

just as claimed at the end of chapter 8.

Note the power of the diagram. Not only does it enable us to lay out clearly on the page the rather tricky situation in which Bob is looking at Alice, while Alice in turn is looking at Bob, but the rules relating intervals to areas of light rectangles contain the quantitative information needed to establish geometrically the relation (10.16), without any further tales of Alice and Bob accompanied by algebraic manipulations.

Figure 10.26 is the very similar construction appropriate when the events E_1 and E_2 are space-like separated. I do not repeat the argument that now the squared interval between the events is

$$I^2 = (t_3 - t_1)(t_1 - t_2), \tag{10.17}$$

because it is virtually identical to the one just given. (Notice that the caption of figure 10.26 is identical to the caption of figure 10.25.)

Since the interval between space-like separated events is the distance between them in the frame in which they happen at the same time, this procedure gives a way to measure *distances* using only a single *clock*. If Alice and her clock are stationary in the frame in which the two events happen at the same time, then it's easy to understand why the procedure works. The beautiful thing is that it continues to work even if she's not.

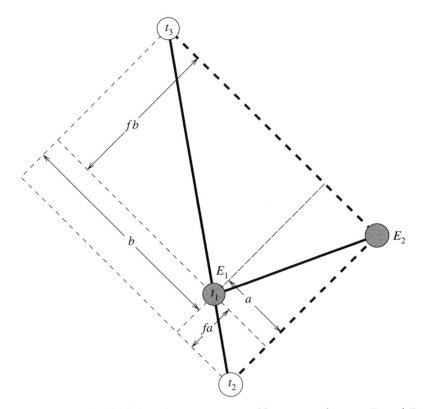

Figure 10.26. The shaded circles are two space-like separated events E_1 and E_2. The heavy solid line is a clock moving uniformly with Alice that is present at E_1 when it reads t_1. The lower heavy dashed photon line demonstrates that somebody present at the event E_2 sees Alice's clock reading t_2 when E_2 takes place. The upper heavy dashed photon line demonstrates that Alice's clock reads t_3 when Alice sees E_2 taking place.

We have shown that the two light rectangles in figure 10.20 have the same area, but how are their shapes related? We can characterize the shape of a light rectangle by specifying the ratio of the lengths of its sides, called its *aspect ratio*. All light rectangles with parallel diagonals have the same aspect ratio, so we can answer the question by comparing any two rectangles whose diagonals are equilocs of Alice and Bob. Such rectangles can be found in figure 10.27, which shows an equiloc for Bob, connecting points P and R, and equilocs and equitemps for Alice, passing through P and through R and intersecting at point Q. Bob's light rectangles have aspect ratio f/F, while Alice's have aspect ratio e/E.

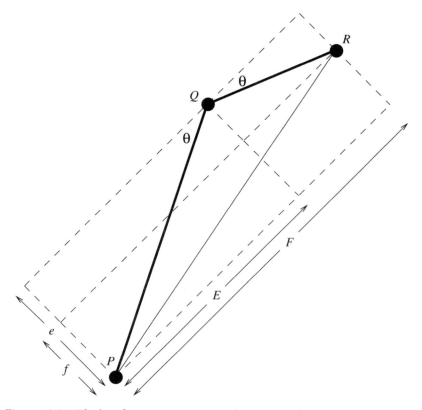

Figure 10.27. The line from P to R is an equiloc in Bob's frame. The heavier lines are Alice's equiloc from P to Q and her equitemp from Q to R. They make the same angle θ with the photon line through Q. Bob's velocity v in Alice's frame is the ratio of the lengths of these two heavy lines. Each of the lines forms the hypotenuse of a right triangle whose other sides are given by various other photon line segments. Because the two angles θ are the same, the right triangles are similar, and the ratios of corresponding sides also give v.

We can relate the two aspect ratios to the velocity v of Bob in Alice's frame by noting that v is the ratio of the length of Alice's equitemp from Q to R to the length of her equiloc from P to Q. Each of these lines is the hypotenuse of a right triangle, and because the lines are equitemps and equilocs for Alice, the two right triangles are similar. Consequently the ratio v of the lengths of the two lines is equal to the ratio of the lengths of either pair of corresponding sides, and we have

$$v = \frac{e-f}{e} = \frac{F-E}{E}. \qquad (10.18)$$

These relations tell us that

$$f/e = 1 - v \ \text{ and } \ F/E = 1 + v. \tag{10.19}$$

Consequently, the aspect ratio f/F of Bob's rectangle is related to the aspect ratio e/E of Alice's by

$$\frac{f/F}{e/E} = \frac{1-v}{1+v}, \tag{10.20}$$

where v is the velocity of Bob in Alice's frame of reference.

With this information, we can extract the quantitative expression for the relativistic Doppler shift from figure 10.21. Applied to the two photon rectangles in that figure, (10.20) tells us that

$$\frac{1-v}{1+v} = \frac{b/B}{a/A}, \tag{10.21}$$

where v is the velocity of Bob in Alice's frame of reference. But the two photon rectangles in figure 10.21 have the same area:

$$Bb = Aa. \tag{10.22}$$

It follows from (10.22) that $b/a = A/B$, so we can rewrite (10.21) as

$$\frac{1-v}{1+v} = (b/a)(A/B) = (b/a)^2 = (A/B)^2. \tag{10.23}$$

But, as (10.4) shows, either of the ratios b/a or A/B are equal to the Doppler factor $f = t/T$, the rate at which Alice or Bob sees time passing on the other's clock, as measured by her or his own clock. Consequently,

$$f = \sqrt{\frac{1-v}{1+v}}, \tag{10.24}$$

where v is their relative velocity in feet per nanosecond, just as we found in chapter 7.

Eleven

$E = Mc^2$

ONE CANNOT WRITE A BOOK on relativity without including a chapter on $E = Mc^2$, the second most famous equation of all time. I would put ahead of it only the discovery of Pythagoras, which we have also had occasion to use, that the area of the square on the hypotenuse of a right triangle is equal to the sum of the areas of the squares on the other two side: $C^2 = A^2 + B^2$. To understand Einstein's celebrated relation between energy (E) and mass (M), we shall have to examine a third quantity, momentum (P). (Why momentum is always denoted by P or p is something of a mystery; but M and m can't be used, since they are reserved for mass.)

We begin our journey to $E = Mc^2$ by examining how mass is defined nonrelativistically. I remind you that "nonrelativistically" does not mean ignoring the principle of relativity; on the contrary, it means maintaining the principle of relativity, but applying it only in cases in which all relevant speeds are very small compared with the speed of light, c. Only after an exploration of the nonrelativistic case will we be in a position to see how to generalize the definition of mass so it can be used to describe objects whose speeds are not small compared with c. We shall see that the nonrelativistic definition of mass survives intact in the relativistic case, with one small proviso.

Although it may violate principles of good pedagogy, I begin with two ways *not* to define mass. The first bad definition of mass is due to Newton, who defined mass to be "quantity of matter." This is useless for at least two reasons. How can you count up the quantity of matter in something? If matter was all built out of identical little bricks, one might be able to do it by counting the number of bricks, but unfortunately matter (as we understand it today) is made up of many different kinds of bricks, so this doesn't work unless you have an independent definition for the "quantity of matter" in two bricks of different types. Furthermore, even when you put together identical bricks, it turns out (in the relativistic case) that the mass of the object you construct depends on how you put the bricks together.

Here is a second bad definition, whose badness is sometimes belabored to the point of boredom in elementary physics courses: mass is weight. The problem here is that the weight of an object is the force that gravity exerts on it. This depends on where the object is. An object's weight on the Moon is about a sixth of its weight on Earth (which itself varies a bit depending on where you are on Earth), and its weight in empty space is zero. But its mass is the same whether it is on Earth, on the Moon, or adrift in interstellar space.

Here, finally, is a qualitative formulation of the correct definition: The mass of an object is a measure of how hard it resists attempts to change its velocity. Under a given set of circumstances, the bigger the mass of an object, the less its velocity changes. This is far too informal a statement to stand by itself, but it captures the essence of the concept. It's easy to push around a beachball, harder to push around a solid wooden ball of the same size, and extremely hard to push around a solid lead ball of that size. By defining mass in terms of changes in velocities, so that measuring a mass reduces to measuring times and distances in an appropriate experiment, the correct definition is ideally suited for a reexamination in the relativistic case.

To make the concept of mass less qualitative, we must go beyond this informal definition and state some simple facts, which make it possible to give quite a precise definition. To do this we return to the kinds of collisions we were considering in chapter 1, in which two particles come together with certain velocities, collide, and go off with certain other velocities. We must now take particular care to distinguish explicitly between velocity and speed. I shall follow the widespread practice of using bold face letters ($\mathbf{u}$, $\mathbf{v}$, $\mathbf{w}$) for velocities , which can be positive or negative, and italic letters (u, v, w) for speeds, which are always positive. Thus a particle with speed u has velocity $\mathbf{u} = u$ if it moves east along the tracks, and $\mathbf{u} = -u$ if it moves west. Note that the square of a velocity is the same as the square of the corresponding speed: $\mathbf{u}^2 = u^2$.

The correct and precise definition of mass is contained in one crucial fact about such collisions. It is possible to associate with every particle a positive number m, called its *mass*, which is a measure of how little the velocity of the particle changes in such collisions; the bigger the mass, the less the change in velocity. To give the precise relation between the masses of the two particles and their changes in velocity, call the particles 1 and 2, and their masses m_1 and m_2. Call their velocities before the collision $\mathbf{u}_1^b$ and $\mathbf{u}_2^b$, and their velocities after the collision $\mathbf{u}_1^a$ and $\mathbf{u}_2^a$, so that the *changes* in their velocities are $\mathbf{u}_1^a - \mathbf{u}_1^b$ and $\mathbf{u}_2^a - \mathbf{u}_2^b$. (Note that b and a now stand for "before" and "after," and not for "Bob" and "Alice.") Many experiments establish the extremely important fact that the comparative size of the two changes in velocity is entirely determined by the comparative size of the two masses, according to the simple rule:

$$\frac{\mathbf{u}_1^a - \mathbf{u}_1^b}{\mathbf{u}_2^a - \mathbf{u}_2^b} = -\frac{m_2}{m_1}. \tag{11.1}$$

Since the masses are positive, the minus sign simply means that the ratio of the changes in velocities is negative—i.e. the change in the velocity of one of the particles is in the opposite direction from the change in velocity of the other. If the velocity of one increases, the velocity of the other decreases. Note that increasing or decreasing *velocity* is not the same as increasing or decreasing *speed*. If a particle moving to the east slows down a little, its velocity decreases. But if a particle moving to the west speeds up, its velocity also decreases, because it becomes a larger negative number. And if a particle moving to the west slows down a little, its velocity increases, because it becomes a smaller negative number.

The content of (11.1) is simple and intuitive. If two particles collide and experience equal but opposite changes in velocity, then (11.1) tells us they have the same mass. If the change in velocity of one particle is only half the size of that of the other, then the particle with the smaller change in velocity has twice the mass of the other particle. If one particle suffers a change in velocity only 1 percent the size of that experienced by the other, then it must be 100 times more massive.

In the nonrelativistic theory the rule embodied in (11.1) holds whatever the individual velocities happen to be. One might expect there to be trouble when speeds approach significant fractions of the speed of light, and indeed the rule then fails to hold, as we shall see. However even in the more accurate relativistic theory, as one also might expect—indeed, as one must require—the rule holds to a very high degree of precision provided all particle speeds are small compared with the speed of light. This requirement makes it possible to use the nonrelativistic definition of mass emerging from (11.1) to define mass even in the relativistic theory. One simply makes the additional proviso that all the particle speeds in a collision designed to compare the masses of two particles must be small compared with the speed of light. "How small?" you might ask. That depends on how accurately you want to know the ratio of the masses. Since no mass is known to better than about 10 significant figures, an error of one part in 10 billion is good enough for all practical purposes, which, as we shall see, means the speeds ought to be less than a hundred thousandth of the speed of light, about 10 feet per millisecond—roughly 10 times the speed of sound in air and therefore still a pretty brisk clip.

Implicit in the definition (11.1) of mass is the fact that the same number m works for a given particle regardless of what other particle it collides with. Thus although our definition gives only the comparative resistance to changes of velocity of a pair of particles, we end up with the same collection of masses for all particles regardless of which pairs we choose to test against each other. For example by testing particles 1 and 2 we learn the ratio m_2/m_1, and by testing particles 2 and 3 we learn m_3/m_2. The product of these two ratios is m_3/m_1, and indeed, if we test particles

1 and 3 directly this is precisely what we get. Note that there is nothing in the nature of collision experiments that logically requires that this should be so. It is an important fact about nature that different kinds of particles behave in this very simple way when they collide at nonrelativistic speeds.

Of course we can only determine in this way the *ratio* of the masses of all the particles. The overall scale is arbitrary and can be fixed, for example, by taking one standard object and declaring its mass to be "1 kilogram."

It is very important that this nonrelativistic definition of mass is consistent with the principle of relativity. The numbers you get for the mass ratios do not depend on the frame of reference in which the collision is described, *provided* we use the *nonrelativistic* velocity addition law. For if we view all the collisions in a frame moving to the right with speed v— i.e., with a velocity **v** that is positive—then every velocity **u** appearing in (11.1) is replaced by $\mathbf{u} - \mathbf{v}$, which leaves all *changes* in velocity unaffected, and these are the only things appearing in (11.1).

Here and in what follows, you should be careful not to confuse the relative velocity **v** of two frames of reference with the velocities **u** of the particles participating in a collision: **v** is fixed throughout the collision and has nothing to do with the collision itself. It is merely the relative velocity of two frames whose descriptions of the collision we are interested in comparing. The individual particle velocities **u**, on the other hand, can vary from one particle to another and will in general change in the course of a collision.

That this nonrelativistic definition of mass works in any inertial frame of reference is, of course, crucial if it is to be viewed as embodying a law of nature, for the principle of relativity requires laws of nature to hold in all inertial frames. This indicates already that something goes awry in the relativistic case, for the relativistic rule is that when you change frames of reference **u** is replaced by $\frac{\mathbf{u}-\mathbf{v}}{1-\mathbf{uv}/c^2}$. Of course if both speeds u and v are small compared with the speed of light c, this difference is so small as to be unimportant. But if **u** and **v** are both comparable to c, then changes of velocity can depend significantly on the frame of reference. So the rule (11.1) can *only* be valid in the nonrelativistic case.

This nonrelativistic definition of mass leads quite naturally to the non-relativistic definition of momentum. With a little elementary algebraic manipulation, we can rewrite (11.1) in the mathematically equivalent form:

$$m_1\mathbf{u}_1^b + m_2\mathbf{u}_2^b = m_1\mathbf{u}_1^a + m_2\mathbf{u}_2^a. \tag{11.2}$$

Although this has precisely the same mathematical content as (11.1), it presents the information in a different way. The left side of (11.2) only contains velocities before the collision, whereas the right side only contains velocities after. We have therefore discovered a quantity that is unchanged, or *conserved*, by the collision. It is called the total momentum, usually denoted by the symbol **P**. We call it "total" momentum because it is convenient also to define the momentum **p** of an individual particle of mass m and velocity u by

$$\mathbf{p} = m\mathbf{u}, \tag{11.3}$$

so that the total momentum **P** of two particles is just

$$\mathbf{P} = \mathbf{p}_1 + \mathbf{p}_2. \tag{11.4}$$

Equation (11.2) is the nonrelativistic law of conservation of momentum. From our point of view, it is just a reformulation of our definition of mass. But like that "definition," it has profound physical content going well beyond merely a conventional definition. It is a remarkable *fact* that it is *possible* to assign to every particle a number m in such a way that momentum—the weighted sum of the particle velocities, in which the velocity of each particle is weighted by its own mass—is indeed conserved in all nonrelativistic collisions between all possible pairs of particles.

Conservation of momentum holds under very general nonrelativistic conditions. It continues to hold when more than two particles participate in the collision. It also continues to hold even when the motion of the particles is not confined to a single line. In that case one must specify the velocity of a particle by its components along three different directions—for example, up-down velocity, north-south velocity, and east-west velocity. The generalized law then says that momentum is independently conserved for each of these three different components. Conservation of momentum even continues to hold when the numbers or kinds of particles *change* as a result of their collision.

Suppose, for example, particles 1 and 2 stick together to form a single new particle, particle 3. When this happens the mass of particle 3 turns out to be just the sum of the masses of the original two, and momentum continues to be conserved. It is of crucial importance that m_3 should indeed be $m_1 + m_2$. For if all the velocities **u** are replaced by $\mathbf{u} - \mathbf{v}$, then the momentum before the collision is reduced by $(m_1 + m_2)\mathbf{v}$, while the momentum after the collision is reduced by $m_3\mathbf{v}$. Thus if m_3 were not $m_1 + m_2$, momentum would not be conserved in the new frame. This is so important that it is stated as a law of conservation of mass: if two particles m_1 and m_2 merge into a single particle of mass M, then

$$M = m_1 + m_2. \tag{11.5}$$

If the law of conservation of mass did not hold nonrelativistically, then the law of conservation of momentum could not hold either, since momentum could not be conserved in all frames of reference. The meaning of $E = mc^2$ is closely related to the fact that conservation of mass often fails to hold in the relativistic case, as we shall see.

To understand how energy E enters the story, it is useful to examine a two-particle collision, in the special frame of reference in which the total momentum is zero. In this zero-momentum frame—a term preferred by physicists is "center-of-mass frame," but we shall use the more descriptive term—we have before the collision

$$m_1 \mathbf{u}_1^b + m_2 \mathbf{u}_2^b = 0 \tag{11.6}$$

and, because momentum is conserved,

$$m_1 \mathbf{u}_1^a + m_2 \mathbf{u}_2^a = 0 \tag{11.7}$$

after the collision too. In the zero-momentum frame the particles move in opposite directions, since the velocities of 1 and 2 have to have opposite signs if their momenta add up to give zero. So in the zero-momentum frame the particles come together and then fly apart with speeds whose ratios are the same both before and after the collision:

$$\frac{u_2^b}{u_1^b} = \frac{m_1}{m_2} = \frac{u_2^a}{u_1^a}. \tag{11.8}$$

But although the *ratios* of the speeds are the same both before and after the collision, there is nothing in the law of conservation of momentum to require the speeds *individually* to stay the same. Momentum conservation is consistent with both speeds either increasing or decreasing in the zero-momentum frame, as long as the percentage increase or decrease is the same for both particles. There is, however, something special about a collision in which both speeds remain the same—i.e. in which the particles simply bounce back in the directions they came from with their original speeds. One calls such collisions *elastic*, and calls *inelastic* those collisions in which the individual speeds change in the zero-momentum frame. An inelastic collision in which both speeds dropped might be one in which the particles tended to stick together when in contact, and therefore lost some of their speed in the course of pulling apart again. An inelastic collision in which both speeds increased might be one in which a small explosive charge was set off when the particles touched, propelling them back faster than they came together. It is an important fact that momentum continues to be conserved even in cases like these.

Whatever the reason for a collision being elastic or inelastic, one singles out elastic collisions for special treatment in the nonrelativistic theory, because in an elastic collision something else, besides momentum, is conserved. In the zero-momentum frame of two particles, it is the individual speeds themselves that are conserved, but that is special to both the zero-momentum frame and the case of two particles. It is easy to see what the new quantity must be if we require it to be conserved in *all* frames of reference. Define the "kinetic energy" k of a particle of mass m and velocity $\mathbf{u}$ by

$$k = \tfrac{1}{2}m\mathbf{u}^2, \tag{11.9}$$

and define the total kinetic energy of two particles to be

$$K = k_1 + k_2. \tag{11.10}$$

The factor $\tfrac{1}{2}$ is entirely a matter of convention, designed to make things come out simpler further on. (Clearly we could redefine any of the quantities m, p, or k by introducing arbitrary numerical scale factors that were the same for all particles.)

Since u_1 and u_2 are *separately* conserved in an elastic collision in the zero-momentum frame, so are k_1 and k_2 and hence their sum. Any number of other possible definitions of K would share this property. What makes the particular definition (11.9) special is that if K is conserved in *one* frame of reference, then it will necessarily be conserved in *all* frames. This is a matter of some practical importance, because it means that we do not have to transform our description to the zero-momentum frame to check on whether a collision is elastic. We need only compute $K = \tfrac{1}{2}m_1\mathbf{u}_1^2 + \tfrac{1}{2}m_2\mathbf{u}_2^2$ both before and after the collision; the collision is elastic if and only if K is the same before and after.

How does K change when we change frames? The velocity $\mathbf{u}$ changes to $\mathbf{u} - \mathbf{v}$, so the kinetic energy $k = \tfrac{1}{2}m\mathbf{u}^2$ changes to

$$k' = \tfrac{1}{2}m(\mathbf{u} - \mathbf{v})^2 = \tfrac{1}{2}m\mathbf{u}^2 - m\mathbf{u}\mathbf{v} + \tfrac{1}{2}m\mathbf{v}^2 = k - \mathbf{p}\mathbf{v} + \tfrac{1}{2}m\mathbf{v}^2. \tag{11.11}$$

If we have two or more particles, we just add up the changes in kinetic energy for each of them, so the total kinetic energy in the new frame is

$$K' = K - \mathbf{P}\mathbf{v} + \tfrac{1}{2}M\mathbf{v}^2, \tag{11.12}$$

where $\mathbf{P}$ is the total momentum and M is the total mass. Suppose the total kinetic energy in the zero-momentum frame K is the same before and after the collision. Then since the total momentum $\mathbf{P}$ and the total mass M are also conserved, it follows that the kinetic energy K' in the new frame will also be the same before and after the collision. Thus it is a consequence of the conservation of total momentum and total mass, that if total kinetic energy is conserved in one frame, it will be conserved in all frames. If we define a collision to be elastic if kinetic energy is conserved in the collision, then whether or not a collision is elastic is independent of frame of reference.

This concludes our review of the nonrelativistic state of affairs. To summarize, in the nonrelativistic theory mass, momentum, and energy have the following properties:

Mass

We associate with each particle a mass m, a number characteristic of the particle, independent of the frame of reference in which the particle is described; the total mass M of a collection of particles is just the sum of their individual masses. Total mass is conserved in all collisions. Total mass is the same in all frames of reference: if M is the total mass in one frame and M' is the total mass in a frame moving with velocity $\mathbf{v}$, then

$$M' = M. \tag{11.13}$$

Momentum

If a particle of mass m has a velocity $\mathbf{u}$ we define its momentum $\mathbf{p}$ by

$$\mathbf{p} = m\mathbf{u}. \tag{11.14}$$

The total momentum $\mathbf{P}$ of a collection of particles is just the sum of their individual momenta. The total momentum is conserved in all collisions. The momentum $\mathbf{P}'$ in a frame moving with velocity $\mathbf{v}$ is related to the momentum $\mathbf{P}$ in the original frame by

$$\mathbf{P}' = \mathbf{P} - M\mathbf{v}. \tag{11.15}$$

where M is the total mass.

Energy

If a particle of mass m has a velocity $\mathbf{u}$ we define its kinetic energy k by

$$k = \tfrac{1}{2}m\mathbf{u}^2. \tag{11.16}$$

The total kinetic energy K of a collection of particles is just the sum of their individual kinetic energies. The total kinetic energy is only conserved in a special kind of collision, known as an elastic collision. If a collision is elastic in one frame of reference, then it is elastic in all frames. This follows from the fact that the total kinetic energy K' in a frame moving with velocity $\mathbf{v}$ is related to the total kinetic energy K in the original frame by

$$K' = K - \mathbf{Pv} + \tfrac{1}{2}M\mathbf{v}^2, \qquad (11.17)$$

where M is the total mass and $\mathbf{P}$ is the total momentum in the original frame.

There is an important interplay here between conservation laws (which specify quantities that are the same before and after the collision) and transformation rules (which tell how quantities change from one frame of reference to another). A conservation law relates the value of a quantity before the collision to its value after the collision, when both values are computed in the same frame of reference. For it to be a *law*, it must be valid in all frames of reference, so we must use the transformation rules to check that a candidate for a conservation law is capable of being obeyed in all frames of reference. In the case of mass conservation that's easy, since mass is the same in all frames of reference. Momentum can be conserved in all frames of reference because it obeys the transformation rule (11.15) *and* because total mass is the same before and after a collision. Kinetic energy can be conserved in all frames of reference (if it is conserved in any one frame) because it obeys the transformation rule (11.17) *and* because *both* total momentum *and* total mass are the same before and after a collision.

It is also important that the *contingently* conserved quantity, K, does not appear in the transformation rules governing the quantities $\mathbf{P}$ and M that are *always* conserved. If K did appear in the transformation rules for either $\mathbf{P}$ (or M), then since K is not always conserved, neither could $\mathbf{P}$ (or M) always be conserved.

This is the nonrelativistic view of mass, momentum, kinetic energy, and their conservation. But when we get to speeds comparable to the speed of light c, this whole nonrelativistic picture falls apart. The pleasing compatibility of these conservation laws and their ability to be satisfied in all frames of reference makes critical use of the nonrelativistic velocity addition law, $\mathbf{u}' = \mathbf{u} - \mathbf{v}$. When this rule is significantly violated, as it is in the full relativistic description, then conservation of momentum ceases to be valid in all frames if it holds in any one, because the simple transformation rule (11.15) for momentum is no longer valid. The same problem arises with kinetic energy.

This should not come as a surprise. There is no reason to expect that the appropriate forms for the momentum and kinetic energy of a particle should be identical to the forms they have in the nonrelativistic case. After all, not even the rate of a moving clock or the length of a moving stick is the same as in the nonrelativistic case. The question is whether it is possible to find new conservation laws involving suitable generalizations of the nonrelativistic definitions of mass, momentum, and kinetic energy. We are guided in the search for such generalizations by the requirement they have two essential features:

1. They must reduce to the nonrelativistic forms when the speeds of the particles are small compared with the speed of light, since we know the nonrelativistic conservation laws hold to a high degree of accuracy in that limit.

2. If the appropriately generalized quantities are conserved in one frame of reference, then they must be conserved in all frames of reference.

The proper relativistic definition of mass is the easiest to deal with. We retain exactly the same definition of mass as in the nonrelativistic theory, only adding the proviso that the velocities of all particles in a collision used to determine their masses should be small compared with the velocity of light. How small depends on how accurately we want to determine the masses. A good practical criterion is to say that they should be so small that if we repeat the experiment with even smaller velocities, we get exactly the same set of masses to within the accuracy of whatever method we use to determine the relevant speeds. You may protest that this procedure does us no good in determining the mass of a photon, since photons in empty space cannot move at any speed other than the speed of light. We shall take up the special case of photons at the end of this chapter.

As so defined, the mass of a particle continues to be an inherent property of the particle, having nothing to do with how fast the particle might be moving in other collisions in which it might subsequently find itself. It is an invariant, independent of frame of reference. If there were a particle whose mass were not invariant, then we could distinguish one inertial frame from another by performing in each frame a low-velocity collision that determined the mass of the particle. (In the early days of relativity, it was sometimes the practice to give a different relativistic definition of mass that made the mass of a particle depend on its velocity. Compensating changes were made in the relativistic definitions of energy and momentum so that those expressions were the same as those we shall now construct. Today, however, the mass of a particle is always defined to be independent of its velocity.)

We defer for the moment the question of whether total mass, defined as the sum of the masses of the individual particles, continues to be conserved in collisions that change the numbers and types of particles. Note, though, that any failure of mass conservation had better be by a very small amount when the speeds of all particles participating in the collision are small compared with the speed of light c, since the nonrelavistic theory, in which total mass *is* conserved, holds to a high degree of precision when all speeds are small compared with c.

We turn next to the relativistic definition of the momentum of a particle of mass m. Since m continues to be simply an invariant number, characterizing the particle, the question is what quantity can play the role of the particle's velocity $\mathbf{u}$. We have two criteria to meet: (a) the new quantity must reduce to $\mathbf{u}$ when u is small compared with c; (b) when one changes the frame of reference, the new quantity must change in a manner that has a simplicity comparable to the nonrelativistic rule $\mathbf{u}' = \mathbf{u} - \mathbf{v}$, if we are to have a hope of conserving momentum in all frames of reference. The velocity $\mathbf{u}$ itself will not do, for under a change of frame $\mathbf{u}$ changes by the relativistic law:

$$\mathbf{u}' = \frac{\mathbf{u} - \mathbf{v}}{1 - \mathbf{uv}/c^2}. \tag{11.18}$$

Here $\mathbf{u}'$ is the velocity of the particle in the new frame, $\mathbf{u}$ is the velocity of the particle in the old frame, and $\mathbf{v}$ is the velocity of the new frame in the old frame.

It is the denominator in (11.18) that keeps the transformed total momentum $\mathbf{P}' = m_1\mathbf{u}'_1 + m_2\mathbf{u}'_2$ from having a form simple enough to ensure momentum conservation in the new frame. Because $\mathbf{u}$ appears in the denominator (as well as the numerator) of (11.18), if we continue to use the nonrelativistic definition (11.14) of momentum, but use the relativistic transformation law (11.18), then the total momentum in the new frame of reference will depend in detail on the individual velocities of all the particles in the old frame. (In contrast, when we use the nonrelativistic velocity addition law (11.15), the total nonrelativistic momentum in the new frame depends on those individual particle velocities only through that particular combination of velocities which is the total momentum in the old frame.)

Why does relativity introduce this complicated denominator into (11.18)? Think back to the definition of velocity: distance traveled divided by the time it takes. In the nonrelativistic case, changing frames changes the distance traveled, but it does not change the time it takes, so only the numerator changes. In the relativistic case, *both* quantities change when you change frames, resulting in the more elaborate rule (11.18).

This immediately suggests a simple and elegant way out of the problem. For the purpose of defining relativistic momentum, we should generalize the notion of the velocity of a particle to the distance traveled by the particle divided by the time it takes to go that distance, where that time is measured in a special frame that all observers can agree on.

What could that special frame be? To ask the question is to answer it: each particle singles out one and only one special frame of reference—the frame in which the particle is at rest.

Suppose, then, we define a generalized velocity **w** to be the distance a particle travels in a given time with the proviso that this time should always be measured by a clock traveling with the particle. This meets the crucial requirement that **w** is indistinguishable from the ordinary velocity **u** when the speed of the particle is small compared with the speed of light, since a clock moving with the particle then runs slowly by an imperceptibly small amount. Now, however, as we go from one frame to another, the distance going into the definition of **w** changes, but the time does not. So if we redefine the momentum **p** to be m**w**, which makes a negligible difference at nonrelativistic speeds, we might hope to find a simple transformation rule for **p**. Since m is invariant, we need only inquire how **w** transforms.

The generalized velocity **w** differs from the ordinary velocity **u** only because the motion of the particle is timed by a clock that moves with the particle, rather than by clocks that are stationary and synchronized in the frame in which the particle is moving. This moving clock runs *slowly* compared with the stationary clocks, so it will indicate that the particle took *less* time to cover a given distance. The slowing-down factor $\sqrt{1 - u^2/c^2}$ gives the reduction in time, so **w** will be *bigger* than the ordinary velocity **u** by precisely the factor $1/\sqrt{1 - u^2/c^2}$:

$$\mathbf{w} = \mathbf{u}/\sqrt{1 - u^2/c^2}. \tag{11.19}$$

With the expression (11.19) at hand, we can use the relativistic transformation rule (11.18) to find how **w** changes when we change to a frame moving with velocity **v**. In the new frame **w**′ is given by

$$\mathbf{w}' = \mathbf{u}'/\sqrt{1 - u'^2/c^2}, \tag{11.20}$$

where **u**′ is related to **u** by the velocity addition law (11.18). If you substitute (11.18) into (11.20) and simplify the resulting expression, you will find that

$$\mathbf{w}' = \frac{\mathbf{u} - \mathbf{v}}{\sqrt{1 - v^2/c^2}\sqrt{1 - u^2/c^2}}. \tag{11.21}$$

Checking this result is the only slightly messy piece of algebra in this whole business, but the conclusions it leads to are so profound that everybody should suffer through it at least once in a lifetime. If you have not the stomach to do the algebra, at least note that (11.21) is obviously correct when $\mathbf{v} = 0$ (in which case it reduces to $\mathbf{w'} = \mathbf{w}$), when $\mathbf{v} = \mathbf{u}$ (in which case we have gone to a frame in which the velocity of the particle is 0), and when $\mathbf{u} = 0$ (in which case the particle was originally at rest and therefore has velocity $-\mathbf{v}$ in the new frame).

So if we define relativistic momentum by

$$\mathbf{p} = m\mathbf{w} = \frac{m\mathbf{u}}{\sqrt{1 - u^2/c^2}}, \tag{11.22}$$

then (11.21) tells us that

$$\mathbf{p'} = \frac{\mathbf{p} - p^0\mathbf{v}}{\sqrt{1 - v^2/c^2}}, \tag{11.23}$$

where I have defined a new quantity p^0 by

$$p^0 = \frac{m}{\sqrt{1 - u^2/c^2}}. \tag{11.24}$$

This is close to what we want, for the momentum in the new frame is now very simply related to the momentum in the old frame. The factor $\sqrt{1 - v^2/c^2}$ in the denominator in (11.23) may not strike you as so simple, but remember that it is only a number, determined by the relative velocity of the two frames. It is independent of the speed of the particle itself, and therefore exactly the same number will appear in the rule that gives the momentum in the new frame for every single particle participating in the collision.

The only real problem—and it is a serious one—is that something new now appears in the transformation rule (11.23): p^0. To see what this interloper p^0 might signify, let us first consider what happens when the speed u of the particle is small compared with the speed of light. In that case (11.24) tells us that p^0 differs indistinguishably from the mass m. If we replace p^0 by m in the transformation law (11.23) and apply it to the total momentum of a pair of particles, we get

$$\mathbf{P'} = \frac{\mathbf{P} - M\mathbf{v}}{\sqrt{1 - v^2/c^2}}, \tag{11.25}$$

which except for the denominator is just the familiar nonrelativistic transformation law. The denominator is harmless since it is just a fixed number, involving only the relative velocity of the two frames, that remains the same before and after the collision. So we can conclude from (11.25), just as we did in the nonrelativistic case, that if **P** is the same before and after a collision then **P'** will be too, provided the total mass M is conserved in the collision.

But if a particle is not moving at a speed small compared with c, then p^0 is *not* nearly its mass m. If we want to insure that momentum, as defined by (11.22), is conserved in all frames of reference, then we must replace the law of conservation of total mass by a new law: conservation of total p^0. Such a replacement is in keeping with the spirit of our attempted generalization of the nonrelativistic momentum conservation law, for total p^0 is given by

$$P^0 = p_1^0 + p_2^0 = \frac{m_1}{\sqrt{1 - \mathbf{u}_1^2/c^2}} + \frac{m_2}{\sqrt{1 - \mathbf{u}_2^2/c^2}}. \qquad (11.26)$$

Since this reduces to total mass when both velocities are small compared with c, we are discovering that the nonrelativistic law of mass conservation is a limiting case of a more general relativistic law, just as the nonrelativistic law of conservation of total $m\mathbf{u}$ is a limiting case of conservation of a more general relativistic concept of momentum,

$$\mathbf{P} = \mathbf{p}_1 + \mathbf{p}_2 = \frac{m_1\mathbf{u}_1}{\sqrt{1 - \mathbf{u}_1^2/c^2}} + \frac{m_2\mathbf{u}_2}{\sqrt{1 - \mathbf{u}_2^2/c^2}}. \qquad (11.27)$$

But before we can declare there to be a new conservation law for P^0, we must check to see whether it too passes the crucial requirement that a genuine law must be valid in all frames of reference. This leads us to one more unpleasant computation very much like the one that led us to (11.23). We must apply the relativistic velocity addition law (11.18) to the definition

$$p^{0\prime} = \frac{m}{\sqrt{1 - u'^2/c^2}} \qquad (11.28)$$

to express $p^{0\prime}$ in terms of quantities in the original frame. When this is done we find:

$$p^{0\prime} = \frac{p^0 - \mathbf{p}\mathbf{v}/c^2}{\sqrt{1 - v^2/c^2}}. \qquad (11.29)$$

(You can extract the result (11.29) more deftly by dividing the left side of the momentum transformation law (11.23) by the left side of the velocity transformation law (11.18) and the right side by the right side, and comparing what you get with the definitions of p^0 and **p**.)

This has a structure very similar to the transformation rule (11.23) for momentum. Because both structures are so simple, the transformations (11.23) and (11.29) for the individual particle $\mathbf{p}$ and p^0 lead to transformations for the total momentum, $\mathbf{P}$, and total p^0, P_0, of *exactly* the same forms as (11.23) and (11.29):

$$\mathbf{P}' = \frac{\mathbf{P} - P^0\mathbf{v}}{\sqrt{1 - v^2/c^2}}, \tag{11.30}$$

$$P^{0'} = \frac{P^0 - \mathbf{Pv}/c^2}{\sqrt{1 - v^2/c^2}}. \tag{11.31}$$

Since these express $\mathbf{P}'$ and $P^{0'}$ entirely in terms of $\mathbf{P}$ and P^0 (and the relative velocity $\mathbf{v}$ of the two frames) if the unprimed quantities are the same before and after a collision, the primed quantities must be too. Therefore if $\mathbf{P}$ and P^0 are both conserved in one frame, they will both be conserved in any other frame. Our proposed relativistic generalization (11.22) of the definition of momentum meets our criteria for a conserved quantity, as does the new quantity P^0, whose conservation we are forced to consider.

What are the implications of replacing the nonrelativistic conservation of total mass M by the relativistic conservation of P^0? How are we to interpret p^0 and the sum P^0 of the values of p^0 for a group of particles? We can get a powerful clue by examining the structure of p^0 for a particle whose speed u is small compared with c. In this limit the definition (11.24) merely tells us what we already know: that p^0 is very close to m, the mass of the particle. But since we are trying to make sense of the difference between the old nonrelativistic law of conservation of M and a new relativistic law of conservation of p^0, what we really require is an estimate of the *difference* between p^0 and m, when u is small compared with c. We would like that estimate to be a little more informative than the declaration that the difference is very small. A little later in this chapter (pp. 163–64) we construct such an estimate, showing that when u is very small compared to c, then to a high degree of accuracy,

$$p^0 - m = \tfrac{1}{2}mu^2/c^2. \tag{11.32}$$

Thus at nonrelativistic velocities $p^0 - m$ *is nothing but the nonrelativistic kinetic energy divided by* c^2. (The analysis leading to (11.32) is not complicated, but I can't bear to interrupt the narrative at this exciting moment.)

So if we define the *relativistic kinetic energy* by

$$k = p^0 c^2 - mc^2, \qquad (11.33)$$

then k does indeed reduce to the ordinary nonrelativistic kinetic energy at speeds small compared to c, and we have our interpretation of p^0: the interesting quantity is not p^0 itself, but the product of p^0 with c^2, which is the sum of two terms:

$$p^0 c^2 = mc^2 + k. \qquad (11.34)$$

We have now reached our goal. In order for the total relativistic momentum $\mathbf{P}$ to be conserved, it is necessary for P^0 to be conserved as well, where

$$P^0 c^2 = Mc^2 + K, \qquad (11.35)$$

M and K being the total mass and total relativistic kinetic energy.

Recall now the nonrelativistic state of affairs. Total mass M is always conserved, but total kinetic energy K is conserved only in elastic collisions. Relativistically we can continue to define elastic collisions as those in which total relativistic kinetic energy K is conserved. But relativistically P^0 must always be conserved, whether or not the collision is elastic, for if it were not, momentum could not be conserved in all frames. Since P^0 is related to M and K through (11.35), it follows that if K is conserved, then M must be conserved as well. But if K is not conserved, then M cannot be conserved either. In an inelastic collision, if the total kinetic energy changes by $\Delta K = K^a - K^b$, then in order for P^0 to be conserved in the collision, (11.35) requires this change in kinetic energy to be precisely balanced by a change in the total mass by $\Delta M = M^b - M^a$:

$$\Delta Mc^2 = \Delta K. \qquad (11.36)$$

This balancing of a loss (or gain) in kinetic energy by a compensating gain (or loss) in mass must be true whether the collision involves relativistic or nonrelativistic velocities, since the relativistic theory ought to be valid for all velocities. Why, then did we never notice it in collisions at nonrelativistic velocities, where total mass appeared to be conserved? The reason is that at nonrelativistic velocities the change in mass is just too small to be measurable. This change is $\Delta M = \Delta K / c^2$, and a measure of the size of ΔK, the change of kinetic energy, is the total mass M times the square of a typical particle velocity u^2. Thus the change in ΔM is typically the mass M itself times a factor whose size is roughly u^2 / c^2. At less than supersonic velocities, u^2/c^2 is less than $1/1,000,000,000,000$. But no mass has ever been measured to a precision of one part in a trillion.

So the change in mass required in inelastic collisions by the relativistic theory is far too small to be observed in collisions at nonrelativstic speeds. The exact relativistic conservation of p^0c^2 simply masquerades as conservation of total mass when all speeds are small compared with the speed of light. But at relativistic speeds, the consequences of the correct relativistic conservation law can be substantial.

Returning from the sublime to the merely conventional, I note that one defines P^0c^2 to be E, the total energy, and defines p^0c^2 for each individual particle to be its energy e. One then has the energy and momentum of a particle of mass m and velocity u defined by

$$e = \frac{mc^2}{\sqrt{1 - u^2/c^2}}, \tag{11.37}$$

$$\mathbf{p} = \frac{m\mathbf{u}}{\sqrt{1 - u^2/c^2}}. \tag{11.38}$$

The transformation rules (11.30) and (11.31) become

$$E' = \frac{E - \mathbf{P}v}{\sqrt{1 - v^2/c^2}}, \tag{11.39}$$

$$\mathbf{P}' = \frac{\mathbf{P} - E\mathbf{v}/c^2}{\sqrt{1 - v^2/c^2}}. \tag{11.40}$$

Note that (11.37) asserts that the energy e of a particle of mass m has the value mc^2 when the particle is at rest. This is sometimes incorrectly cited to be the meaning of $E = Mc^2$. The true meaning is to be found in inelastic collisions, as the expression (11.36) of the necessity for a change in mass to compensate exactly for any change in kinetic energy: if kinetic energy is gained in the collision, total mass must go down; if kinetic energy is lost, total mass must go up.

If, for example, two objects collide in their zero-momentum frame and stick together to form a final object at rest, the mass of that final object will exceed the sum of the masses of the two colliding objects by precisely their kinetic energy prior to the collision divided by c^2. Physicists wishing to create new "elementary" particles, significantly more massive than any that have been observed to date, must therefore fling together less massive particles at speeds comparable to c, in order to provide the kinetic energy needed to supply the additional post-collision mass. Whence the popularity of particle accelerators among these scientists.

Conversely, if a particle at rest spontaneously explodes into two particles that go flying apart, the total mass of the final two particles must be less than the mass of their parent by precisely their kinetic energy divided by c^2. It is often said that the awesome power of a nuclear explosion is a direct manifestation of $E = Mc^2$. This is no more or less correct than to say that an ordinary chemical explosion is such a manifestation. In both cases the total mass of everything flying off in the explosion is less than the mass of the original ingredients of the explosion by the total kinetic energy produced in the explosion divided by c^2. The difference is that in a chemical explosion things fly apart at speeds that are tiny on the scale of the speed of light, so the change in mass would be too small to detect, even if you could collect everything flying away from the explosion. But the forces responsible for a nuclear explosion are very much stronger than chemical forces, and as a result things fly apart at speeds so large that the change in mass can be as big as 0.1 percent of the original mass—still a small fraction, but no longer too small to measure.

So it is more to the point to say that a nuclear explosion is substantially more powerful than an ordinary chemical explosion because the nuclear forces that play a role in such explosions are substantially stronger than the electrical forces that play a role in chemical bonding and hence in ordinary explosions. That the nuclear forces are so strong that one can actually detect the (still small but not too small to measure) change in mass is an impressive manifestation of the strength of those forces.

One of the earliest manifestations of the fact that the forces holding atomic nuclei together were indeed extraordinarily strong came from the fact that the masses of nuclei differed by a few parts in a thousand from the masses of the constituents out of which they were built. In his very first paper on $E = Mc^2$, written just a few months after his first relativity paper, Einstein suggested that this could account for the discrepancy: "It is not excluded that a test of the theory can be achieved with bodies whose energy content is variable to a high degree (e.g. radium salts)."[1]

The fact that the prodigious energy released in a nuclear explosion is still only about $\frac{1}{1000}$ of the mass of the explosive provides a dramatic measure of the difficulty of achieving (or stopping) motion comparable to the speed of light. Suppose, for example, an object of mass m moves at $\frac{3}{5}$ the speed of light, so its energy (11.37) is $e = \frac{5}{4}mc^2$, and therefore its kinetic energy is $k = e - mc^2 = \frac{1}{4}mc^2$. Setting off a nuclear warhead of

---

[1] From A. Einstein, "Does the Inertia of a Body Depend on its Energy Content", *Annalen der Physik* **18** (1905): 639–41.

mass M produces an energy of about $\frac{1}{1000}Mc^2$, so to get an object moving at $\frac{3}{5}$ the speed of light requires the energy of a nuclear warhead 250 times as massive as the object.

To put it another way, if an object moving at $\frac{3}{5}$ the speed of light is somehow brought to an abrupt halt, say by colliding with a spectacularly immobile and impenetrable barrier, then the energy released in the collision will be that of a nuclear warhead 250 times as massive as the object. Rocket travel near light speed will not be as easy as it is in the movies.

But although the relativistic definition (11.37) of energy implies that the energy of a particle grows without bound as its speed gets closer and closer to that of light, there are nevertheless particles (the photon, for example) that do move at the speed of light. Since it does not take infinite energy to produce a photon, how are we to account for this?

The first thing to notice is that (11.37) does allow a particle to move at a speed u equal to the speed of light c without having an infinite energy, provided the mass of the particle is zero. But it would appear that the relativistic definitions of energy and momentum, (11.37) and (11.38), can tell us nothing useful about zero mass particles with speeds $u = c$, since dividing zero by zero is a famous way of arriving at utter nonsense. There are, however, two consequences of these two equations that remain well defined even in the limit of zero m.

It easily follows from (11.37) and (11.38) that

$$e^2 = p^2c^2 + m^2c^4, \tag{11.41}$$

and that

$$\mathbf{p} = e\mathbf{u}/c^2. \tag{11.42}$$

You can also go the other way around: starting with (11.42) and (11.41) you can easily deduce (11.37) and (11.38). The two pairs of equations are completely equivalent. But, unlike (11.37) and (11.38), the equivalent forms (11.42) and (11.41) retain a perfectly intelligible content even when applied to particles of zero mass. When $m = 0$, (11.41) reduces to

$$p = e/c. \tag{11.43}$$

The relation (11.42) then holds, provided the speed u of the zero mass particle is equal to the invariant speed c.

Thus the relativistic definitions of energy and momentum can be applied to a particle of zero mass, provided one takes them to require that the speed of such a particle is necessarily the speed of light c, and that the energy of such a particle is just c times the magnitude of its momentum.

It turns out that for most purposes (11.41) and (11.42) are much easier to work with than (11.37) and (11.38), even for particles with nonzero mass, so that while (11.37) and (11.38) play a fundamental role in motivating the new definitions of energy and momentum, it is (11.41) and (11.42) that capture their features most effectively.

You can view the quantity p^0 from a perspective that ties together the concepts of relativistic energy and momentum in a way that is simply unavailable in the nonrelativistic case. The momentum of a particle in any given frame of reference is the product of the mass of the particle with the rate at which the particle moves through *space* as measured by a clock moving with the particle. In quite the same way, p^0 is the mass of the particle times the rate at which the particle moves through *time*, as measured by a clock moving with the particle.

This sounds crazy to nonrelativistic ears: how can something move through time at anything but a rate of 1 second per second. And indeed, nonrelativistically conservation of p^0 is just conservation of mass. But relativistically it makes perfect sense as a rather elegant way to express the slowing down of a moving clock. The higher the speed of a particle (in a given frame of reference), the more rapidly the particle moves through time (as measured in that frame) according to a clock moving with the particle (which measures time in its proper frame—called *proper time*). Thus in a frame in which a particle moves at $\frac{3}{5}$ the speed of light, it moves through time at a rate of $\frac{5}{4}$ of a second per proper second. This is just a dramatic, upside-down and, in some deep sense, more meaningful way of saying that any internal clock-like processes associated with the particle run slowly by the appropriate slowing-down factor: for every second that passes on any clock moving with the particle, time in the frame in which we are describing this motion advances by 1.25 seconds.

When something speeds up its passage through space, so that it takes it less proper time to get from here to there, it also speeds up its passage through time, so that it takes it less proper time to get from now to then. (At the end of chapter 8, I remarked that as a clock speeds up its passage through space, its passage through time slows down. The apparent contradiction is reconciled by the fact that there we measured the passage of a moving clock through time, in a given frame, by the rate at which time advanced on the clock, per second of time in that frame; but here we are talking about the rate at which time passes, in the given frame, per second of time as given by the moving clock.)

There remains a piece of unfinished business: we must justify using the form (11.32) for $p^0 - m$ when the speed u of a particle is small compared with the speed of light c. It follows from the definitions of $\mathbf{p}$ and p^0 (11.22) and (11.24) that

$$p^{0^2} - \mathbf{p}^2/c^2 = m^2 \tag{11.44}$$

or

$$p^{0^2} - m^2 = (p^0 - m)(p^0 + m) = \mathbf{p}^2/c^2 \tag{11.45}$$

or

$$(p^0 - m) = \frac{\mathbf{p}^2}{(p^0 + m)c^2}. \tag{11.46}$$

The left side of (11.46) is what we are looking for: the difference between p^0 and m. The right side unfortunately contains p^0 again, but if we are only interested in speeds u small compared with c, then p^0 is exceedingly close to m. When u is small compared with c, we can evaluate the right side of (11.46) with very high accuracy by replacing the p^0 on the right with m. Under these same conditions, $\mathbf{p}$ is also exceedingly close to the nonrelativistic value $m\mathbf{u}$. Making both these replacements on the right side of (11.46) gives us the estimate we are looking for. When a particle moves slowly compared with the speed of light, to a high degree of precision,

$$p^0 - m = \tfrac{1}{2}mu^2/c^2. \tag{11.47}$$

Table 11.1 summarizes all the features of mass, momentum, and energy that we have uncovered in both the nonrelativistic and relativistic cases, in a form that directly compares the two cases.

We conclude with some illustrations of how the relativistic conservation laws can be used in some simple collision problems, not unlike those we examined in chapter 1.

As an illustration of how the laws work in an extreme relativistic case, consider a collision between a photon (which moves, of course, at the extremely relativistic speed c) and an initially stationary particle of mass m_i. Suppose the photon is absorbed by the particle. This is a relativistic (and asymmetric—the two particles are no longer identical) version of the collision we examined in chapter 1 between two particles that stick together and form a single compound particle after the collision. If the photon has energy ω, how fast does the particle move after it has absorbed the photon, and what is the particle's new mass m_f? (The subscripts i and f stand for "initial" and "final"—a convenient alternative to "before"

TABLE 11.1 Comparison of the relativistic and nonrelativistic properties of energy, momentum, and mass of a pair of particles

	Nonrelativistic	Relativistic
Mass	$M = m_1 + m_2$	$M = m_1 + m_2$
Conserved?	always	elastic collisions only
Transformation	$M' = M$	$M' = M$
Momentum	$P = m_1 u_1 + m_2 u_2$	$P = \dfrac{m_1 u_1}{\sqrt{1-u_1^2/c^2}} + \dfrac{m_2 u_2}{\sqrt{1-u_2^2/c^2}}$
Conserved?	always	always
Transformation	$P' = P - Mv$	$P' = \dfrac{P - vE/c^2}{\sqrt{1-v^2/c^2}}$
Energy	$E = \frac{1}{2}m_1 u_1^2 + \frac{1}{2}m_2 u_2^2$	$E = \dfrac{m_1 c^2}{\sqrt{1-u_1^2/c^2}} + \dfrac{m_2 c^2}{\sqrt{1-u_2^2/c^2}}$
Conserved?	elastic collisions only	always
Transformation	$E' = E - Pv + \frac{1}{2}Mv^2$	$E' = \dfrac{E - vP}{\sqrt{1-v^2/c^2}}$

Note: "Conserved" means that the quantity is the same before and after the collision. The entries under "transformation" give with a prime ($'$) the values the quantities have in a frame moving with velocity **v** with respect to a frame in which they have the values given without primes. The conservation laws obey the principle of relativity: if they hold in one inertial frame then they hold in all inertial frames. But this is true for the nonrelativistic quantities only if one uses the (in general incorrect) nonrelativistic velocity addition law for changing frames of reference. To compare the relativistic and nonrelativistic states of affairs it is useful to note that when the speed of a particle is small compared with the speed of light, then its energy, $\frac{mc^2}{\sqrt{1-u^2/c^2}}$ is very nearly equal to $mc^2 + \frac{1}{2}mu^2$. Note the different roles played by inelastic collisions in the relativistic and nonrelativistic theories. Nonrelativistically, mass is conserved even in inelastic collisions but kinetic energy is not; relativistically, energy is conserved even in inelastic collisions but mass is not. Although the entries in the table refer to a pair of particles, the same relations hold for any number of particles. The number of particles before and after the collision need not be the same. If there is only one particle after the collision, then we are describing several particles fusing into one; if there is only one particle before the "collision," then we are describing a particle that disintegrates into more than one.

and "after" when applying the conservation laws.) The answers to these questions fall directly out of the conservation laws for total energy and momentum.

Before the collision the photon has energy ω and the particle has energy $m_i c^2$, since this is what (11.37)—or (11.41) and (11.42) together—give for a particle with mass m_i and speed $u = 0$. After the collision the particle has swallowed up the photon and has energy e. Conservation of energy requires:

$$\omega + m_i c^2 = e. \tag{11.48}$$

Before the collision the photon has momentum k, which (11.43) tells us is related to its energy ω by

$$k = \omega/c, \tag{11.49}$$

and the particle has momentum 0, since it is stationary. After the collision the particle has momentum p and there is no photon left. So conservation of total momentum requires the final particle to have all the momentum initially possessed by the photon:

$$\omega/c = p. \tag{11.50}$$

If you know the energy and the momentum of an object, you can most easily extract its velocity directly from (11.42): if something moves with velocity $\mathbf{u}$, its energy e and momentum $\mathbf{p}$ are related by $\mathbf{p} = e\mathbf{u}/c^2$. Therefore the ratio of its speed to the speed of light is given by

$$u/c = cp/e. \tag{11.51}$$

Using the forms (11.50) and (11.48) for p and e gives the answer:

$$u/c = \frac{1}{1 + m_i c^2/\omega}. \tag{11.52}$$

If $m_i c^2$ is large compared with the energy ω of the photon, then the speed of the particle after the collision is a small fraction of the speed of light. But when the energy ω of the photon becomes comparable to $m_i c^2$ of the particle, the speed with which the particle recoils becomes comparable to c. To get the particle moving at speeds very close to the speed of light c, the energy ω of the photon must become much larger than $m_i c^2$. Note, though, that no matter how large ω becomes, (11.52) still gives a final speed u for the particle that is less than the speed of light.

The simplest way to learn the mass m_f of the particle after it has absorbed the photon is through the relation (11.41) between the energy, momentum, and mass of a particle. Applied to the particle after it has absorbed the photon, this gives

$$(m_f c^2)^2 = e^2 - (pc)^2. \tag{11.53}$$

Using the forms (11.48) and (11.50) for e and p, we learn from (11.53) that m_f satisfies

$$(m_f c^2)^2 = (\omega + m_i c^2)^2 - \omega^2 = (m_i c^2)(2\omega + m_i c^2), \tag{11.54}$$

so the mass of the particle after it has absorbed the photon has become

$$m_f = m_i \sqrt{1 + 2\omega/m_i c^2}. \tag{11.55}$$

Thus, the mass m_f of the particle after it has absorbed the photon can be significantly larger than its original mass m_i, provided the energy ω of the photon is comparable to or exceeds $m_i c^2$.

Note that one can use (11.52) to reexpress the relation (11.55) between the initial and final masses in terms of the velocity u of the final particle. The result of this small calculation is the curious fact that the ratio of the masses is given by nothing but the Doppler shift factor from chapter 7:

$$m_f/m_i = \sqrt{\frac{1 + u/c}{1 - u/c}}. \tag{11.56}$$

We can also give a relativistic treatment of the precise problem discussed in chapter 1, in which two identical objects stick together when they collide. In the frame in which they are fired directly at each other with equal speeds, the symmetry of the situation requires the resulting compound object to be stationary, even in the relativistic case. But what happens if one is initially stationary and the other is fired directly at it?

In the nonrelativistic case, it was easy to find a frame in which the new situation reduced to the old one: the frame moving in the direction of the first object with half its speed u. This led to the conclusion that in the original frame the compound object would move at a speed of $\frac{1}{2}u$. We can solve the problem the same way in the relativistic case, but finding the velocity of the frame in which the collision is symmetric is more complicated, for we must use the relativistic velocity addition law. If we view the collision in a frame moving in the same direction as the first object with speed v, then in that frame the second object, stationary in the original frame, moves in the opposite direction with speed v. The speed of the first object in the new frame is

$$w = \frac{u - v}{1 - uv}. \tag{11.57}$$

I have chosen here to measure distances in feet and times in nanoseconds, so since $c = 1$ foot per nanosecond we can leave it out of the addition law, making things a lot simpler. This is a trick that professional users of relativity exploit all the time. All velocities that emerge at the end will be in feet per nanosecond and can therefore be directly interpreted as fractions of the velocity of light. If you like, you can reintroduce c at the end by replacing every velocity v by the fraction v/c.

For the collision to be symmetric in the moving frame, we need the speed w of the first object in that frame also to be v, and therefore we must find a speed v that satisfies

$$v = \frac{u - v}{1 - uv}. \tag{11.58}$$

The speed in the original frame of the final compound object will be v in the original frame, since it is stationary in the frame in which the collision is symmetric. The condition (11.58) leads to a quadratic equation for the unknown velocity v, which, while not beyond our powers to cope with, is more complicated than anything algebraic that has come up in this book, even in the simplified form it assumes when the formula is uncluttered by factors of c.

It is easier to find the answer directly in the original frame, by using the conservation laws. The final velocity u_f of the compound object is related to its momentum p_f and energy e_f by (11.42), which (with $c = 1$ f/ns) gives

$$u_f = p_f/e_f. \tag{11.59}$$

Since there is only one object after the collision, p_f is the final total momentum. Before the collision only the first object is moving, so the initial total momentum is the momentum p_i of that moving object. Conservation of total momentum therefore gives

$$p_f = p_i. \tag{11.60}$$

The initial total energy is the energy e_i of the moving object and the energy m of the stationary object. (Its energy is just its mass m because its velocity is zero and because in units of feet and nanoseconds $c^2 = 1$.) So energy conservation gives

$$e_f = e_i + m. \tag{11.61}$$

We can use these two conservation laws to reexpress the speed u_f in (11.59) in terms of quantities before the collision:

$$u_f = \frac{p_i}{e_i + m} = \frac{e_i u_i}{e_i + m} = \frac{u_i}{1 + m/e_i}. \tag{11.62}$$

But applying (11.37) (with $c = 1$) to m/e_i gives the desired relation between initial and final speeds:

$$u_f = \frac{u_i}{1 + \sqrt{1 - u_i^2}}. \tag{11.63}$$

When u_i is small compared with 1 f/ns, (11.63) gets very close to the nonrelativistic answer $u_f = \frac{1}{2}u_i$, as it must. But when the first object is hurled at the stationary object with a speed u_i close to 1 f/ns, (11.63) tells us that the speed u_f of the compound object is hardly less than u_i. An object moving at close to the speed of light is very hard to slow down. If you try, by putting another identical object in its path, it just picks it up and carries it along, with hardly a change of speed.

As noted, we could have deduced (11.63), though with considerably more effort, by the old trick of changing frames used in chapter 1. What we cannot determine by that method is the mass M of the resulting compound object. This, however, follows from (11.41) (again simplified by the fact that $c = 1$ f/ns):

$$M^2 = e_f^2 - p_f^2 = (e_i + m)^2 - p_i^2, \tag{11.64}$$

the second form following from using conservation of energy and momentum to replace the final total energy and momentum by their expressions in terms of initial quantities. A little algebraic manipulation then gives

$$M^2 = e_i^2 + 2e_i m + m^2 - p_i^2. \tag{11.65}$$

But (11.41) tells us that $e_i^2 - p_i^2 = m^2$, so (11.65) simplifies to

$$M^2 = 2m(e_i + m). \tag{11.66}$$

It is convenient to replace e_i by $m + k$, where k is the kinetic energy of the moving object. We then have

$$M = 2m\sqrt{1 + k/2m}. \qquad (11.67)$$

In the nonrelativistic case u_i is a small fraction of the speed of light, so $k = \frac{1}{2}mu_i^2$ is a tiny fraction of m, and (11.67) gives us the expected nonrelativistic answer that the mass M of the compound object is just twice the mass m of the two identical objects that collided to produce it. In the "extreme relativistic" case, where the initial speed of the first object is very close to the speed of light, so its kinetic energy k is much larger than m (times c^2, but $c = 1$ f/ns), then 1 can be ignored compared with $k/2m$, and (11.67) gives

$$M = \sqrt{2km}. \qquad (11.68)$$

This is a result of some practical importance. It says that if you want to produce a single particle of large mass M by firing one particle of small mass m against another stationary particle of mass m, then the mass M you can produce grows only as the square root of the kinetic energy of the incident particle. If you want to increase M by a factor of 10, you have to increase k by a factor of 100. If, on the other hand, you try to do the same thing by firing two particles of mass m at each other with equal speeds, then momentum conservation ensures that the final composite particle of mass M will be at rest, and all the kinetic energy of the colliding particles can be converted into its mass. To increase M by a factor of 10, you only have to increase the kinetic energies of the particles by the same factor of 10. This is why particle accelerators are designed to fire particles at each other—so-called colliding beams. This advantage of colliding beams over firing particles at a stationary target only emerges when the speeds are very close to the speed of light. It is a manifestation of relativity at the level of engineering.

Twelve _____

A Bit about General Relativity

As NOTED IN CHAPTER 2, Einstein was led to his extraordinary insights into the nature of time by his conviction that the laws of electromagnetism ought to be consistent with the principle of relativity, just as the laws of mechanics are. Electric and magnetic forces were not, however, the only forces of fundamental interest in 1905. There was also the force of gravity. Einstein spent another 10 years trying to extend relativity to gravitational phenomena. He succeeded in 1915 with his *general theory of relativity*, which has become of fundamental importance in cosmology, in astrophysics, and even—remarkably for a subject that was long thought to be of only intellectual interest—in the very practical matter of how the global positioning system (GPS) operates here on Earth. Unlike special relativity, general relativity defies presentation at the thorough but mathematically elementary level of this book. But at least one of Einstein's early insights fits in quite well. This is his discovery (of crucial importance for the GPS) that gravity affects the rate at which a clock runs, a small though important piece of the puzzle, achieved before his complete formulation of 1915.

The effect of gravity on the rate of a clock emerges from a new principle, enunciated by Einstein quite early in his search for a relativistic theory of gravity. He called it the *principle of equivalence*, and it is as general and powerful as the two principles that underly special relativity. The principle of equivalence was suggested to Einstein by the fact, discovered by Galileo, who here played yet another role at the foundations of relativity, that the effect of gravity on the motion of an object is independent of its mass.

In a vacuum, where there is no air resistance, a light wooden ball takes the same time to fall to the ground from a given height as a massive lead ball. Near the surface of the Earth both of them undergo uniform acceleration by the same amount—each second their downward velocities increase by a little over 30 feet per second. This had been known for such a long time that people tended to take it entirely for granted. But if you look at it in the right way, as Einstein was wise enough to do, it is far from obvious. The mass of an object, as emphasized in chapter 11, is a measure of how hard it is to change its velocity. Why should gravity act more strenuously to change the velocity of more massive objects in just such a way as to lead to exactly the same changes in velocity, whatever the mass?

An immediate consequence of this curious behavior of gravity is that if Alice is falling from a height with a lot of other objects like her namesake in Wonderland, and she observes the objects falling with her, they will all appear to her to be stationary or moving uniformly, as if she were stationary in an inertial frame of reference and gravity were not acting at all. The same effect is demonstrated by the behavior of objects inside the international space station. Although the Earth's gravity at the space station is almost as strong as it is at the surface of the Earth, objects within the space station behave in its frame of reference as if the Earth's gravity were absent.

You might object that this is a bad example. The space station maintains its height above the surface of the Earth, so it is not falling. This objection, however, overlooks the curvature of the Earth. The space station is indeed falling toward the Earth all the time. But it has also been given such a large horizontal velocity that, as it falls, the surface of the round Earth beneath it curves away from it by just the same amount that it has fallen. As a result, it can continue falling indefinitely, without ever getting closer to the ground. As it falls, the ground falls away from it, so it falls, as it were, *around* the Earth, as illustrated in figure 12.1. It is this condition, that the arc of a satellite's trajectory must match the curvature of the Earth, that in fact determines how large (about 35,000 fps) its horizontal velocity must be. Indeed, almost all of the energy required to put something into low Earth orbit is devoted not to lifting it to the height of its orbit, but to producing the kinetic energy associated with the horizontal velocity needed to maintain that orbit.

The direction in which things accelerate under the influence of the Earth's gravity is not fixed; it is directed toward the center of the Earth. And the magnitude of that acceleration depends on the distance from the center of the Earth, being weaker, for example, at the orbit of the moon than it is near the Earth's surface. But if one is only interested in a region that is small on the scale of the whole planet— for example the city of New York and the space extending above it for a few kilometers—then the magnitude and direction of the gravitational acceleration hardly vary at all. One says that the gravitational field is uniform.

We shall ignore the complexity of nonuniform gravitational fields here, confining our explorations to regions in which the magnitude and direction of the gravitationally induced acceleration does not vary. Much of the mathematical complexity of general relativity results from the need to apply the principle of equivalence in a manner that stitches together little patches in which gravity is almost exactly uniform, in order to deal with nonuniform gravitational fields such as that prevailing over the entire Earth. Another important component of the theory that we shall not address is how a gravitational field is determined by the distribution of matter that gives rise to it.

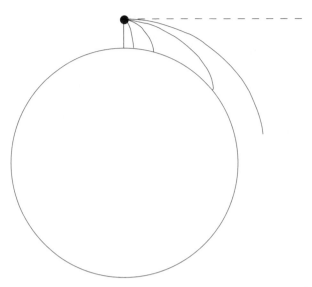

Figure 12.1. (After Newton.) The large white circle represents the Earth. The small black circle represents a launching platform, held stationary above the Earth by rocket engines. If an object is dropped from the platform, it falls directly to the Earth along the vertical line. The curving lines to the right of the vertical line represent the trajectories of other objects that are launched from the platform with greater and greater initial horizontal velocities. If the horizontal velocity is great enough, as it is for the last of these lines, although the object still falls, it never gets closer to the surface of the Earth. That it is nevertheless falling—i.e., it is acted on by the Earth's gravity—is demonstrated by the dashed horizontal line, which is the trajectory it would have taken had the Earth's gravity not acted on it at all.

The rule given by the principle of equivalence, then, is this: Within a uniform gravitational field—i.e. within a region in which gravity acts in the same ("downward") direction, imposing the same constant acceleration g on all objects throughout their fall—all effects of this gravitational field on all physical phenomena are indistinguishable from how the phenomena would play out in the absence of any gravity, but in a frame of reference that is uniformly accelerating (with respect to an intertial frame) in the opposite ("upward") direction with an acceleration a whose magnitude is the same as g. Remarkably, one can conclude directly from this that gravity must affect the rate at which a clock runs. Examining this will occupy us for the rest of this chapter. It will be our glimpse into general relativity.

Suppose we have two identical clocks separated by a fixed distance D. Let the line directed from one clock to the other define the direction we call "vertical," so we can refer to the clocks as the "upper" clock and the "lower" clock, whether or not gravity acts to produce a downward acceleration along that line. Suppose the upper clock emits f_U (f for "frequency"), little pulses of light between each of its ticks, and the pulses move (at the speed of light, of course) down to the lower clock. It's helpful to think of f_U as a very large number (say a million pulses per tick), so the time between pulses indicated on the upper clock is just a millionth of a tick. We want to know how many pulses the lower clock receives from the upper clock for each of its own ticks. Call this f_L.

If both clocks are stationary in an inertial frame in the absence of gravity, the answer is clear. Each pulse covers the same distance and therefore takes the same time to get from the upper clock to the lower one, so the lower clock receives pulses at the same rate f that the upper clock emits them. Furthermore (and importantly!), since both clocks are identical and stationary in the same inertial frame of reference, their ticks provide identical equally reliable measures of time. We can therefore use the upper clock to time the rate at which pulses leave the upper clock: $f = f_U$. And we can use the lower clock to measure the rate at which pulses arrive at the lower clock: $f = f_L$. Therefore $f_U = f_L$.

The preceding argument might appear to be belaboring the obvious. But consider now the more interesting and less obvious case in which both clocks are stationary in a uniform gravitational field, with the upper clock directly above the lower one. The upper clock could, for example, be at the top and the lower clock at the bottom of a vertical tower on the surface of the Earth. We would like to know whether the presence of gravity alters the relation between f_U and f_L and, if it does, what this implies about the comparative rates at which the two clocks are running.

One's first reaction is that this can't make any difference. We are still in a steady state. Each pulse emitted by the upper clock is received by the lower clock, and there is no accumulation of pulses between the clocks. So the lower clock must still be receiving pulses at the same rate that the upper clock is emitting them. This is entirely correct. The issue, however, is what this implies for the rates f_U and f_L, where f_U is measured by one clock, and f_L, by the other. These rates need not be the same, if the upper clock runs at a different rate from the lower in the presence of gravity.

We can find the relation between f_U and f_L in the presence of gravity by appealing to the principle of equivalence. We know that the relation for stationary clocks in a gravitational field must be the same as the relation for clocks that are uniformly accelerating in the vertical direction in the absence of a gravitational field. So we have to go back to empty space, in the absence of gravity, and examine a situation in which both clocks are accelerating in the "vertical" direction. The lower clock moves faster and faster toward the upper clock, while the upper moves faster and faster away from the lower. We describe all this in an inertial frame with respect to which the two clocks are undergoing this acceleration.

Since the clocks are accelerating, in the inertial frame they will have velocities u equal to gt, where g is the acceleration and t is the time since they were initially at rest. Consequently you might worry about the slowing-down factor $\sqrt{1 - u^2/c^2}$. But we are going to consider a situation in which both clocks start at rest and their velocities never have time enough to grow to anything that isn't tiny compared with the speed of light. Now if u/c is a very small fraction, then $(u/c)^2$ is a very small fraction of a very small fraction—i.e. a very, very small fraction. Since the gravitational effect we're going to extract involves differences in clock rates of a size u/c, the comparatively tiny differences of a size $(u/c)^2$ are inconsequential and can be ignored. We can, in fact, treat this situation nonrelativistically!

Since the clocks are a distance D apart, it takes a time $t = D/c$ for each pulse from the upper clock to reach the lower clock. More accurately, it takes a little less time than that, since the lower clock is moving toward the point where the pulse was emitted. But since the velocity v of the lower clock is tiny compared with the velocity c of each pulse, this difference too is inconsequential.

What *is* of crucial importance is that both clocks have been accelerating during the time of about $t = D/c$ that it takes a group of several successive pulses to get from the upper clock to the lower one. (Let D be large enough and the time between successive pulses small enough so that the time between successive pulses is tiny compared with the time it takes a pulse to travel from one clock to the other.) So by the time the lower clock receives a group of pulses, it is moving faster than the upper clock was when those pulses were emitted by an amount

$$u = gt = g(D/c). \tag{12.1}$$

Because the lower clock is moving toward the pulses with this additional speed u, it collects the pulses at a faster rate than the rate at which they were emitted from the upper clock. This is nothing but the Doppler effect that we examined in considerable detail in chapter 7. The difference between the rate at which the pulses are received by the lower clock and emitted by the upper clock is given by the *nonrelativistic* Doppler factor $1 + \frac{u}{c}$. (The difference between this factor and the relativistic factor we found in chapter 7 is utterly negligible, because of the identity $\sqrt{(1 + \frac{u}{c})/(1 - \frac{u}{c})} = (1 + \frac{u}{c})/\sqrt{1 - (\frac{u}{c})^2}$, and because of the very very small size of $(\frac{u}{c})^2$.)

Now time, in the inertial frame in which we're doing our analysis, is the same as the time indicated by either of the clocks, since they are not moving fast enough for the slowing-down factor to make a difference: the ticks of both the clocks continue to give accurate measures of time in the inertial frame. Since the lower clock is receiving more pulses in a given time than the upper clock emits in the same amount of time, and since the ticks of either clock measure time to a high degree of precision, the lower clock must be receiving more pulses between each of its own ticks than the upper clock emits between each of its own ticks. This difference is given by the Doppler factor $1 + \frac{u}{c}$. We have therefore answered the question about the relation between f_U and f_L when the clocks are accelerating in an inertial frame:

$$f_L = f_U(1 + \tfrac{u}{c}) = f_U(1 + gD/c^2). \tag{12.2}$$

But the principle of equivalence now assures us that the relation (12.2), between the number of pulses f_L the lower clock receives between its successive ticks and the number of pulses f_U the upper clock emits between its own successive ticks, must also hold in a frame of reference in which both clocks are stationary in a uniform gravitational field. If we have two identically constructed clocks, one a distance D above the other in a uniform gravitational field characterized by an acceleration g, then if the upper clock emits f_U pulses of light between each of its ticks, the lower one will receive a larger number of pulses, f_L, between each of its own ticks, where

$$f_L = f_U(1 + gD/c^2). \tag{12.3}$$

But as we have already noted, in a gravitational field the lower clock has to be receiving pulses at the *same* rate as the upper clock is emitting them. This is entirely consistent with the discrepancy (12.3) in the rates f_U and f_L at which the pulses are emitted and received *as measured by ticks of the emitting and receiving clocks,* provided the presence of the gravitational field causes those clocks to tick at different rates. It must be that in a uniform gravitational field the lower clock runs slower than the upper one by precisely the factor $1 + gD/c^2$ appearing in (12.3). This is Einstein's conclusion about the effect of gravity on the rate of a clock.

Note that this gravitational effect on the comparative rates of the two clocks, $(1 + gD/c^2)$, is indeed much more important than the slowing-down factor $\sqrt{1 - (u/c)^2}$ that we ignored in doing our analysis in the inertial frame, since the size of u we needed for the argument wasn't significantly larger than $gt = gD/c$. Thus when $1 + u/c = 1 + gD/c^2 = 1.00001$, $\sqrt{1 - u^2/c^2} = \sqrt{1 - 0.000\,000\,000\,1}$, which is about $1 - 0.000\,000\,000\,05$. This is indeed tiny in comparison.

This effect gives an interesting additional view of the twin paradox, discussed in chapter 10. If Alice makes a high-speed journey away from Earth and back again, she will discover when she gets home that she has aged less than her stay-at-home twin sister Carol, because of the slowing down of all her biological processes required by her motion with respect to Carol, who remains stationary in an inertial frame throughout the process. The "paradox" is that it might appear that Alice could argue that it is Carol who should have aged less, since from Alice's point of view it is Carol who has moved away at high speeds and then returned.

But this attempt to extract a paradox forgets that, unlike Carol, Alice is not stationary in a single inertial frame. She is stationary in the frame of the outgoing space ship during her outward journey and stationary in the different frame of the incoming space ship during her journey back home. Although Carol does age less during both the outward and inward journeys according to the outward and inward frames of reference, there is an additional correction that Alice must make to account for Carol's total aging. This comes from the fact that when she changes ships, she moves from a frame in which the notion of "the time on Earth at this moment" changes by $T = 2Du/c^2$, from Du/c^2 earlier to Du/c^2 later, where u is the speed of each ship and D is the distance from Earth at which the transfer takes place. This additional time is just what is needed to account for the fact that when Carol gets back to Alice, she is actually older than Alice.

Suppose, however, that Alice accomplishes the transition, from leaving Earth at speed u to returning home at the same speed, by reversing the direction of motion of a single spaceship through a process of uniform acceleration in the direction back toward Earth. We can then discuss the entire process by keeping Alice at rest in a *single inertial* frame of reference, provided an appropriate gravitational field is present during the period when Alice and Carol make the transition from moving apart to moving back together. In Alice's frame of reference we must answer two questions: why, in spite of the fact that Alice's engines were firing throughout this period, did she nevertheless remain stationary in an inertial frame; and how did Carol and the Earth where she remained, during this same period, manage to change their velocity from flying away from Alice with speed u to flying back toward her at the same speed.

Both mysteries can be accounted for if a uniform gravitational field appears throughout the turning-around period, with Carol being directly above Alice in that field. Alice's engines were then fired to compensate for the field, maintaining her in her inertial frame by preventing her from being accelerated downward by the gravity. Carol, on the other hand, along with the planet Earth, turned around because nobody made any effort to counteract the effects of this same gravitational field in the vicinity of Earth.

Suppose this wonderful field acts for a time t in Alice's frame of reference. Since Carol's velocity must change by $2u$ (from u away from Alice to u toward her), the size of the uniform acceleration must satisfy $2u = gt$, so

$$g = 2u/t. \tag{12.4}$$

During all the time she is being turned around by gravity Carol is higher in the field than Alice (since the effect of the field is to make her start falling back toward Alice). Consequently, as we have just established, her clocks must be running faster than Alice's by $1 + gD/c^2 = 1 + 2uD/tc^2$. Since the gravitational field is present for the entire time t of the turning around, while Alice ages by t during the turning around, Carol will age by $t(1 + 2uD/tc^2) = t + 2uD/c^2$.

Thus the missing $2uD/c^2$ of aging by Carol now shows up as a gravitational increase of the rate at which Carol and her clocks and everything else on Earth are running during the time the gravitational field acts to turn them all around so they are all heading back to Alice.

This is only a tiny piece of the puzzle solved by the general theory of relativity, though it is one of considerable practical importance for navigating with the global positioning system (GPS), as noted earlier.

Thirteen _____

What Makes It Happen?

IN THE END, YOU MAY BE tempted to regard some of this as intellectual sleight of hand. At the solid, unshakable core of the subject is Einstein's great 1905 discovery that the simultaneity of two events that happen in different places is not an absolute unconditional relation between those events, but a way of talking about them, appropriate to a particular frame of reference, and inappropriate to frames of reference moving with respect to that particular frame along the line joining the events.

This was known long before 1905 for the spatial relation between two events that happen at different times: whether or not they happen in the same place obviously depends on the frame of reference. That it is also true for temporal relations was obscured from when people first started talking about time right up to 1905, because on the human scale of times and distances the amount of time corresponding to a given distance is only 1 nanosecond per foot—too little to notice.

We have seen that Einstein's insight into the conventional nature of simultaneity implies that the rate at which a clock runs and the length of a stick both depend on the frame of reference in which they are measured. In the case of the stick, it's an elementary consequence of the fact that to measure the length of a moving stick you have to note where its two ends are *at the same time*, so if "at the same time" can vary from one frame of reference to another, then so can the length of one and the same stick. In the case of the clock, it's because to learn the rate of a clock you have to note its reading at two different times, so if the clock moves you have to compare it with synchronized stationary clocks in two different places. But if the simultaneity of events in different places is frame-dependent, then so is the synchronization of clocks in different places. So disagreements about simultaneity from one frame to the other necessarily entail disagreements about the rates of moving clocks and the lengths of moving sticks.

But now you may reasonably wonder what *makes* moving sticks shrink, and what *makes* moving clocks run slowly. Do these things really happen? Or are they just secondary manifestations of disagreements about simultaneity leading to disagreements about what constitutes a valid measurement? There is by no means unanimity among practicing physicists on this question, and one frequently finds assertions that, for example, moving clocks *appear to* run slowly when measured by stationary ones, or that moving sticks *appear to* shrink.

But such caution is uncalled for. Moving clocks really do run slowly and moving sticks really do shrink, if the concept of the rate of a clock or the length of a stick is to have any meaning at all. They really must behave in this way, since such behavior follows directly from the $T = Dv/c^2$ rules for simultaneous events and synchronized clocks. It is necessary for clocks and sticks really so to behave if the whole subject is to fit coherently together and not collapse into self-contradiction.

But this is not an entirely satisfactory answer. One yearns for a mechanism. What *causes* moving clocks run slowly? What *causes* moving sticks to shrink? If the only available explanation is that moving clocks run slowly and moving sticks shrink in order to maintain the consistency of relativity, one cannot help wondering why clocks and sticks should know or care about the coherence of relativity. What one really wants is an explanation based on something in the construction of sticks that *requires* them to shrink along the direction of their motion, and something in the mechanism of clocks that *requires* them to slow down when in motion, the shrinking and slowing down both being given by the factor $s = \sqrt{1 - (\frac{u}{c})^2}$.

And indeed, although it can often be quite a complicated matter to spell out in complete detail, one can specify such mechanisms. Within any given frame of reference the physical laws that govern the lengths of sticks and the rates of clocks provide complete and compelling explanations for why a stick must shrink when set into motion along the direction of its length, and why the rate of a clock must diminish when it is set into motion. People who hedge their words with "appear to" have not adequately grasped this fact.

As an example, let us consider an elementary kind of clock which runs slowly when moving for a very simple reason. Take a stationary stick of length D, put a mirror that reflects light back along the stick at each end, and let a pulse of light bounce back and forth along the stick from mirror to mirror, as in part 1 of figure 13.1. The clock ticks every time the pulse makes a complete round trip. Although it might be difficult to make a practical clock that operated in this way, this clock has the virtue of great conceptual simplicity. One needs to know only two facts about the world to analyze how it behaves when moving: how fast the light goes in the moving clock, and how long the stick is when it moves. We know the speed of the light remains c in any inertial frame of reference, so that takes care of the first fact. But you might worry that we must deal with the mechanism causing the shrinking of moving sticks to understand the length of the moving clock. Fortunately there is an easy way to evade this problem: we require the stick not to move parallel to itself but in the perpendicular direction, as shown in part 2 of figure 13.1.

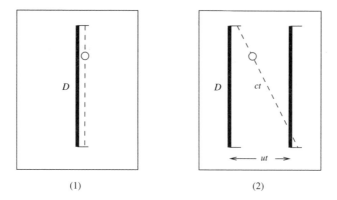

(1)   (2)

Figure 13.1. (1) The heavy line is a stationary stick of length D. The light horizontal lines protruding to the right from its ends are two mirrors. The white circle is a pulse of light that moves back and forth between the mirrors at speed c, along the dashed line. (2) The same stick viewed at two different times from a frame of reference moving with speed u to the left. It takes a time t for the pulse of light to get from the top mirror to the bottom one, during which the stick moves to the right a distance ut. The pulse of light moves along the dashed line. The distance ct traversed by the pulse of light in getting from the top mirror to the bottom one is now $\sqrt{D^2 + (ut)^2}$.

A stick moving *perpendicular* to its length does not shrink (or stretch). The reason is that, in contrast to frames moving along the line joining two events, there is nothing problematic about the simultaneity of events in different places in frames that move *perpendicular* to the line joining the events. The symmetry of the situation makes it impossible to specify which of the events would happen first in the moving frame. They simply have to remain simultaneous. You can also convince yourself that if the events are at opposite ends of a stationary train, then, in contrast to the situation we examined in chapter 5, light originating at the midpoint of the train will reach the two events at the same time in any frame that moves *perpendicular* to the train.

So there is no disagreement, among all the frames moving perpendicular to the stick, about where its two ends are *at the same time*. There is therefore no disagreement about what constitutes a valid measurement of its length. Alice can, for example, simply compare Bob's moving stick with an identical stick stationary in her own frame, also oriented perpendicular to Bob's direction of motion. In both her frame and Bob's all parts of her stick will be adjacent to all parts of Bob's stick *at the same time*. If Alice finds that Bob's stick is shorter than hers, then since Bob agrees that this was a valid measurement, he will conclude that her stick is longer than his. But the principle of relativity prohibits Alice from finding that a stick moving perpendicular to its length shrinks, while Bob finds that it stretches. The principle now requires (as it does not for sticks moving along their own direction, where there is disagreement about what constitutes a valid measurement) that the length of a stick moving perpendicular to itself is unaltered by its motion.

So we can use D for the length of the stick, whether or not it is moving perpendicular to its length. When the clock is stationary it takes a time D/c for light to get from one end to the other, so the time T between ticks is twice this:

$$T = 2D/c. \qquad (13.1)$$

When the clock moves, however, light has to travel a greater distance to get from one end of the stick to the other (see part 2 of Figure 13.1), so because the speed of the light is independent of frame of reference, the clock clearly takes more time between ticks. The mechanism is as simple as that.

Indeed, we can extract a quantitative statement about how much more time it takes. If the speed of the stick is u and the time it takes light to get from one end of the stick to the other is t, then the stick moves a distance ut perpendicular to its length between the departure of the light from one end and its arrival at the other. So the distance the light has to go is the hypotenuse of a right triangle, one side of which has the (unchanged) length D of the stick, and the other side of which has length ut. Since this distance is also the distance ct the light travels in the time t, Pythagoras tells us that

$$(ct)^2 = D^2 + (ut)^2, \qquad (13.2)$$

and therefore

$$t = \frac{D/c}{\sqrt{1 - \left(\frac{u}{c}\right)^2}}. \qquad (13.3)$$

Since the time T_u between ticks of the moving clock is twice this, comparing (13.3) with (13.1) we see that

$$T_u = \frac{T}{\sqrt{1 - \left(\frac{u}{c}\right)^2}}. \qquad (13.4)$$

The moving clock takes longer between ticks than the stationary one, the relation between the two rates being given precisely by the slowing-down factor s that we found, by an entirely different argument, in chapter 6.

A suspicion might remain that this example was picked not for its simplicity, but to exploit the constancy of the velocity of light to provide the sought-for "mechanism" behind the slowing down. This is not a reasonable objection. The constancy of the velocity of light is as fundamental a feature of the world we live in as any of its other physical properties, and a great deal more fundamental than most. It is old-fashioned (by a century) and downright irrational to insist that one not take advantage of it in designing instruments.

Furthermore, a mechanism can always be found, whatever the construction of the clock. It will just be more complicated for most clocks. A modern atomic clock, for example, relies on the frequency of vibration of a certain kind of atom under certain conditions. Such vibration rates are governed by complicated equations of quantum mechanics, first discovered by Paul Dirac. And indeed, if you use the Dirac equation to calculate the frequency of vibration of that same kind of atom under the same conditions *except* that the whole clock is now moving with speed u, you will find, after an enormous effort, that the effect of the motion has been to reduce the rate of vibration by precisely the slowing-down factor $s = \sqrt{1 - (\frac{u}{c})^2}$.

Even the rate of an ordinary old-fashioned mechanical watch is governed by things like the resiliency of certain springs or the inertia of certain wheels. These, in turn, are determined by the forces that hold together the atoms making up the springs and wheels, which are almost exclusively electromagnetic in origin, together with the laws of quantum mechanics that determine the structure of those atoms in the presence of those forces. And although, as far as I know, nobody has ever done a detailed "first principles" calculation of the rate of a moving watch starting at this fundamental level, I can guarantee you that the effect of the motion on the action of those forces will be such as to reduce the rate of the watch by the slowing-down factor.

How can I guarantee it without actually having done the difficult calculation? Recall Einstein's first postulate: the laws of electromagnetism, like those of mechanics, must be consistent with the principle of relativity. One of Einstein's achievements in his 1905 paper was to demonstrate explicitly that those laws could indeed be cast in a form that was explicitly consistent with the principle of relativity. Since the laws also predict that the velocity of light in empty space is independent of the velocity of the source, they are fully consistent with all features of the special theory of relativity. One of those features is that moving clocks run slowly, and therefore any clock whose operation relies on the laws of electromagnetism *must* be required by those laws to run slowly, when set into motion. To fully analyze the operation of the clock one requires, in addition, the laws of quantum mechanics to account for the effect of its motion on the structure of the atoms making up the clock, but one of the constraints Dirac imposed in establishing his equations was precisely that they be consistent with special relativity.

In other words the laws of quantum mechanics (known, more fully, as "relativistic quantum mechanics") and electromagnetism, as we know them today, have been *designed* to guarantee that when applied to calculating the rate of a moving clock, the result emerging from the calculation must be that the moving clock runs slowly. They are also designed to guarantee that the forces holding a stick together behave so as to make the stick shrink when it is set into motion. Those laws must, when applied to such situations, enable one to calculate (though quite possibly with great difficulty and enormous expenditure of computational effort) a detailed physical picture of precisely why moving clocks run slowly and moving sticks shrink.

The fact that the laws of a theory are so designed is often summarized in the assertion that the theory is "Lorentz invariant" or "Lorentz covariant." The terminology honors H. A. Lorentz, who published in 1904 the Lorentz covariant form for the laws of electromagnetism, without, however, understanding the meaning of his achievement, which only emerged with the insight into the nature of time enunciated in 1905 by Einstein.

Is this insistence that any valid physical theory require that moving clocks run slowly and moving sticks shrink a colossal cheat? Do we have mechanistic explanations for these phenomena only because we have refused to consider any fundamental theory that does not provide them? Consider some of the other, more intuitive, invariance principles mentioned in chapter 2. We insist that any acceptable fundamental theory make no distinctions based on absolute orientation. Such theories have rotational invariance built into them. They enable us to give a mechanistic explanation for why somebody can throw a ball just as far to the northeast as he can to the northwest (ignoring wind, the rotation of the earth, and other local irrelevancies). Is this cheating? No! Should we ever actually discover that there was some special direction built into the structure of empty space, it would show up as a failure of one of our fundamental laws, and we would learn something of enormous importance. Since, however, the principle of rotational invariance is correct as far as we know, it seems foolish not to build it into our fundamental formulation of physical theory.

It is the same for the principle of relativity and for the principle of the constancy of the velocity of light, which is better described, in this general context, as the principle of the existence of an invariant velocity, or, if you prefer, the principle of the invariance of the interval. These principles are now so well established that in devising theories to cover new phenomena not embraced by the laws of electromagnetism, we adopt as a guiding rule that these new phenomena should be consistent with the principles on which relativity rests. Since 1905, for example, two new forces have been discovered: the strong interactions responsible for holding together atomic nuclei in spite of the electrical repulsion between the protons making up the nucleus, and the weak interactions that govern certain kinds of radioactive decay, such as the conversion of a neutron into a proton, electron, and neutrino.

These forces have nothing directly to do with electromagnetism, although the weak and electromagnetic forces have now been combined into a single "electro-weak" force that manifests itself in both phenomena. In guessing the laws that govern these forces, powerful constraints were imposed by the invariance principles: that they be consistent with rotational invariance, with translational invariance in space and time, with the principle of relativity, and with the principle of the constancy of the velocity of light. As a result, we are guaranteed that any clock you can design that exploits the strong or weak interactions will, just like an electromagnetic clock, run slowly in a manner given by the slowing-down factor s. And indeed, when it moves at speeds close to c a neutron, which disintegrates in about 12 minutes when at rest, can last considerably longer, thereby serving as a slowly running clock governed primarily by the weak interactions, just like the μ-mesons discussed on p. 65.

So it all hangs together. If "real" has any meaning at all, moving clocks really do run slowly and moving sticks really do shrink. On the other hand, we have come to have a much less naive view of what "real" *really* means. The mechanism that gives the real explanation for a phenomenon in one frame of reference, may be quite different from the mechanism that gives the real explanation in another. Consider, the following example:[1]

Two rockets, are joined by a long rope, stretched tight. At a given moment, both rockets start to move along the direction of the rope (say east) with the same velocity u. Because the rockets start moving at the same time with the same speed, the distance between them does not change. Because they maintain their distance, the need for the now moving rope to shrink is thwarted, and it will be stretched beyond the length that is natural for its new moving condition. If the rockets move fast enough, this stretching will exceed the elastic limit of the rope, and it will break. A fine demonstration of the reality of length contraction.

---

[1] See also the essay by John S. Bell, "How to Teach Special Relativity", in *Speakable and Unspeakable in Quantum Mechanics* (Cambridge: Cambridge University Press, 1987), 67–90.

The story is rather different, however, in a frame moving along the rope with velocity u. Initially both rockets and the (contracted, in this frame) rope are moving west with speed u. Subsequently the eastern rocket stops moving, but the western one continues for a while before stopping, thereby stretching the rope past its breaking point.

Actually both stories are more complicated than this, because the motion imposed on either end of the rope by either rocket is communicated down the rope at only the speed of elastic waves in the rope, which is extraordinarily slowly on the scale of the speed of light, but the more elaborate story also comes in two quite different versions. Each version is absolutely correct—for the frame of reference in which it is told. The simultaneous validity of both stories, relative to the appropriate frame of reference, is no more (or less) strange than the notion people in New York have that people in Sidney, Australia, are all upside down, and vice versa. Both are usefully different ways of talking about the same phenomena.

An important lesson of relativity is that there is less that is intrinsic in things than we once believed. Much of what we used to think was inherent in phenomena turns out to be merely a manifestation of how we choose to talk about them. This is not to say that there is nothing inherent in the phenomena. But that which we have learned is inherent—the interval between two events, for example—turns out to be strange and unfamiliar, while what we once thought was inherent—the time between two events, for example—has turned out to be merely conventional.

This process of discovering that one's former beliefs are wrong, and the painstaking search to identify the old errors, enabling one to construct better founded beliefs to replace them, is what makes the pursuit of science so engrossing. The world would be a far better place for all of us if this joy in exposing one's own misconceptions were more common in other areas of human endeavor.

INDEX